АГ531120

How to Convert Your Car, Van, or Pickup to Diesel

Other TAB books by Paul Dempsey

The Complete Mini-Bike Handbook, No.651

Modern Guide to Auto Tuneup & Emission Control Servicing, No. 688

The Complete Snowmobile Repair Handbook, No. 705

Complete Guide to Outboard Motor Service & Repair, No. 727

Complete Handbook of Lawnmower Repair, No. 767

How to Repair Lift Trucks, No. 796

How to Repair Diesel Engines, No. 817

The Bicycler's Bible, No. 846

Vega, No. 849

How to Repair Small Gasoline Engines, No. 917

Step-By-Step Guide to Ford/Mercury/Lincoln Engine Maintenance/Repair, No. 948

Moped Repair Handbook, No. 976

No. 968
$9.95

How to Convert Your Car, Van, or Pickup to Diesel

By Paul Dempsey

TAB BOOKS
BLUE RIDGE SUMMIT, PA. 17214

FIRST EDITION

FIRST PRINTING—APRIL 1978

Printed in the United States
of America

Library of Congress Cataloging in Publication Data

Dempsey, Paul.
How to convert your car, van, or pickup to diesel.

Includes index.
1. Automobiles—Motors (Diesel). I. Title.
TL229.D5D43 629.2'506 77-18947
ISBN 0-8306-8968-0
ISBN 0-8306-7968-5 pbk.

Cover photo courtesy of Chrysler Corp.

Preface

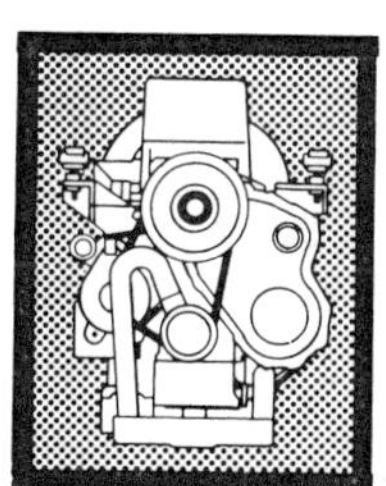

More and more people are turning to diesel power. Diesel engines are the most fuel-efficient known, squeezing more miles from a gallon of lightly refined fuel oil than even the best tuned gasoline plants; and, in spite of the exhaust trails you see behind heavily loaded trucks, compression-ignition engines are clean. The engines discussed here meet amended 1977 emission regulations without emission controls.

Routine maintenance costs are lower for diesels because there are no spark plugs to foul or carburetors to clog. Diesel durability is legendary. Truckers expect a half-million miles between overhauls: until recently, Mack engines were under warranty for 400,000 miles; Detroit engines are warranted for 100,000 miles in city-bus service; and Chrysler engineers have yet to see a Chrysler-Nissan lower end fail while under warranty.

While commercial diesel cars are expensive—you can pay $20,000 for a Mercedes-Benz 300D—it is possible to convert your present vehicle to diesel power at nominal cost. With a new engine, a clean conversion that doesn't run into chassis or steering-gear problems costs less than $3000. Purchasing a used engine and rebuilding it yourself brings the cost down to $1000.

This book, written with the help of C-V Diesel Service of State Line, Maryland, is a guide to making these conversions. One chapter describes the diesel principle, explaining how these engines work and their limitations. Another is devoted to the fuel system, the most critical of all diesel systems and the one requiring most maintenance. Another chapter describes the various engines available, giving dimensions, horsepower, and general technical data on dozens of potential automotive power plants to help you choose the one best for your vehicle. In addition, you will find information here on how to purchase a used engine, what tests to make, and what repairs can be expected. One test described here costs about $105 and will reveal the condition of the engine with scientific certitude.

Another chapter goes into extensive detail on overhaul and rebuild procedures. Some of this material has not been published in a trade book before and is useful to anyone who is elbow-deep in diesel—or gasoline—repair. Another chapter tells you how to choose the vehicle for the swap, what pitfalls to avoid, and something of the nature of the compromises that an engineering project of this magnitude entails. The book concludes with a blow-by-blow description of an engine swap.

Paul Dempsey

Contents

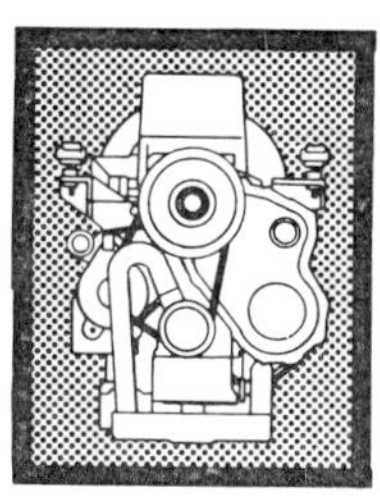

Diesel Power

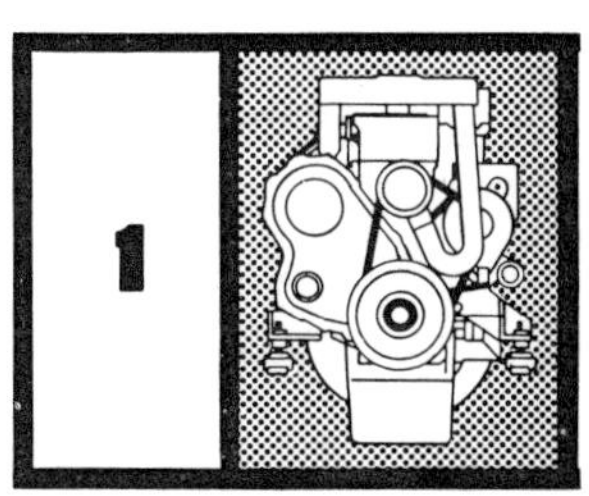

Is it practical to convert your car to diesel power? The answer depends upon a number of factors, some imbedded in the history of the engine, some involving political and economic decisions yet to be made by the federal government.

BACKGROUND AND DEVELOPMENT

In 1892 Dr. Rudolf Diesel filed a proposal with the German patent office to build a compression-ignition engine. This engine promised certain theoretical advantages over steam, particularly in terms of thermal efficiency. Financed by the Krupp family, Diesel constructed a prototype which promptly exploded. A second engine was built and ran under its own power some three years after the original proposal. Series production commenced under licensing agreements with other firms. As in the case of the Wankel engine, each licensee shared the results of his research with the others. Development was rapid, particularly after it was discovered that engine pressures need not remain constant during the power stroke. Augustus Busch, the brewer, purchased the North American rights for $5 million.

Since the diesel engine depends upon the heat of combustion for ignition, fuel must be injected into the cylinder late in the stroke, against compression pressure. Early engines pressurized the fuel charge with air and required two-stage air pumps, a complicated and expensive arrangement. Even so, the diesel was more compact than the steam engines it replaced and twice as fuel efficient. And, as Dr. Diesel was quick to point out, the exhaust was cleaner. Robert Bosch's invention of a high-pressure fuel pump essentially like those used today on Chrysler-Nissan and Mercedes-Benz engines reduced weight and opened the way for relatively small, high-speed engines. Beginning with the French navy, diesels became increasingly important on surface ships before the First World War and the prime power source of submarines. Dr. Diesel did construct a diesel-powered automobile but performance was so disappointing that he put the project aside. In 1921 M.A.N. marketed the first diesel truck, and a few years later, Junkers built a very successful aircraft diesel which developed 900 hp from opposed pistons.

Today the compression-ignition engine is preferred for long-distance trucks, heavy construction equipment, and large ships, the U.S. merchant fleet excepted. This versatile engine is widely used to power generating stations in Europe and to drive irrigation pumps in the Third World. With the exception of oil-starved South Africa and frugal China, most of the world's locomotives are diesel powered.

Mercedes-Benz is usually given credit for the first diesel car to be put in series production. In 1933 Mercedes engineers tested a three-and-a-half liter six-cylinder engine intended to be mounted on the Mannheim chassis. Severe vibration made it impractical for the spindly automobile chassis, and the engineers reluctantly decided to scale the engine down. (Engine swappers often face the same choice between a powerful engine and one that is livable.) In 1936 a four-cylinder version of the Mercedes engine was displayed in the Berlin Automotive Show (Fig. 1-1). Mounted on a 230 chassis with a landaulet body, the engine developed 40 hp at 3000 rpm.

But Mercedes-Benz was not the only firm to exhibit a diesel car that year in Berlin. The Hanomag Rekord-Diesel, a small sedan with an engine having a bore and stroke of 80 by 100 mm. The Hanomag had received preshow publicity at the power plant by setting a racing-car record of 96.8 mph. Although Hanomag records were lost in the war, it appears that about 2500 Rekord-Diesels were built before 1940.

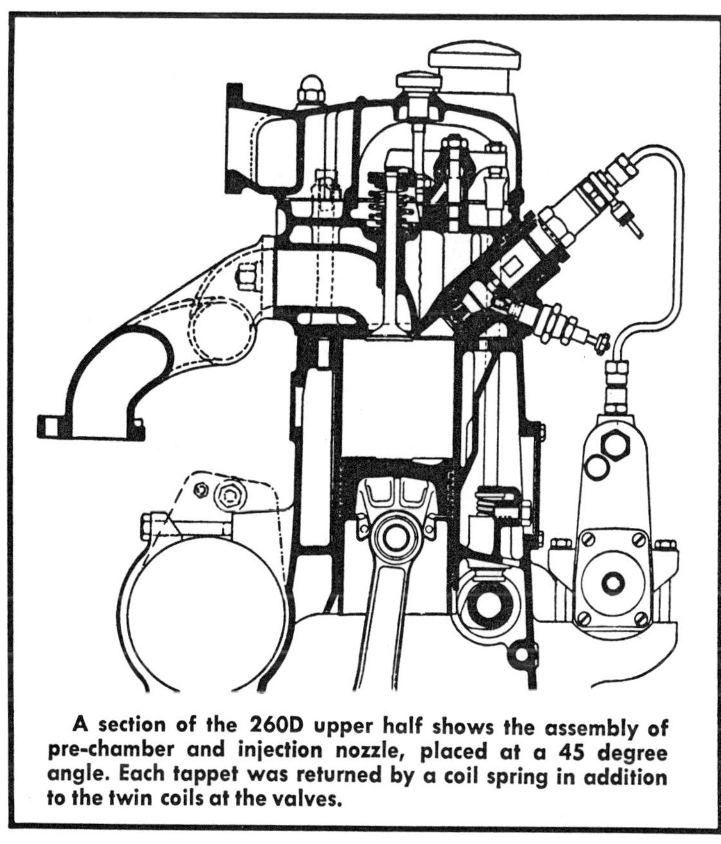

A section of the 260D upper half shows the assembly of pre-chamber and injection nozzle, placed at a 45 degree angle. Each tappet was returned by a coil spring in addition to the twin coils at the valves.

Fig. 1-1. The Mercedes-Benz 260D used spring-loaded tappets, dished pistons, and prechambers.

After the war, Hanomag concentrated on large engines and heavy trucks, leaving automobile production to Mercedes. The first postwar automotive diesel came off the Stuttgart line in the winter of 1949. Its engine was built on the same transfer machines as the 170V gasoline block and shared all its basic dimensions. The four 73.5 by 90 mm cylinders delivered 38 hp. At intervals came the 170DS, the 180D, 190D, 200D*, 220D, 240D, and Mercedes' crowning achievement, the 300D, known in-house as the OM616.

The 300D displaces 3 liters (183.4 cubic inches). It develops 77 hp at 4000 rpm and 115 lb-ft of torque at 2400 rpm. Basically, the

* Not imported into the U.S.

Fig. 1-2. The 300D crankshaft.

300D engine is an expanded version of the 240D, but expanded in a novel fashion: it has an additional cylinder, making it the only five-cylinder auto engine in the world. Each of the five crankshaft throws is 72 degrees apart (Fig. 1-2). For reasons of combustion chamber efficiency it was not practical to increase the size of the four-cylinder's pistons, and a six-cylinder engine would have been too heavy for the chassis. As it is, the five-cylinder's weight of 515 lb is 68 lb more than the four-cylinder's and as much as the 280 gasoline engine available in the same chassis. Despite the 300D's 24 percent increase in displacement over the 240D, EPA figures show the same fuel economy for both: 24 mpg in the city and 31 mpg on the highway.

The 300D engine shares much 240D hardware, including pistons, connecting rods, precombustion chambers, and valve gear (Fig. 1-3). But besides having the additional cylinder and fuel pump plunger, the 300D is different in the shape of its combustion chambers, in its preheater circuitry, and in its governor. The 300D is the first Mercedes to use a centrifugal governor, a much more sophisticated device than the vacuum governor fitted to earlier engines.

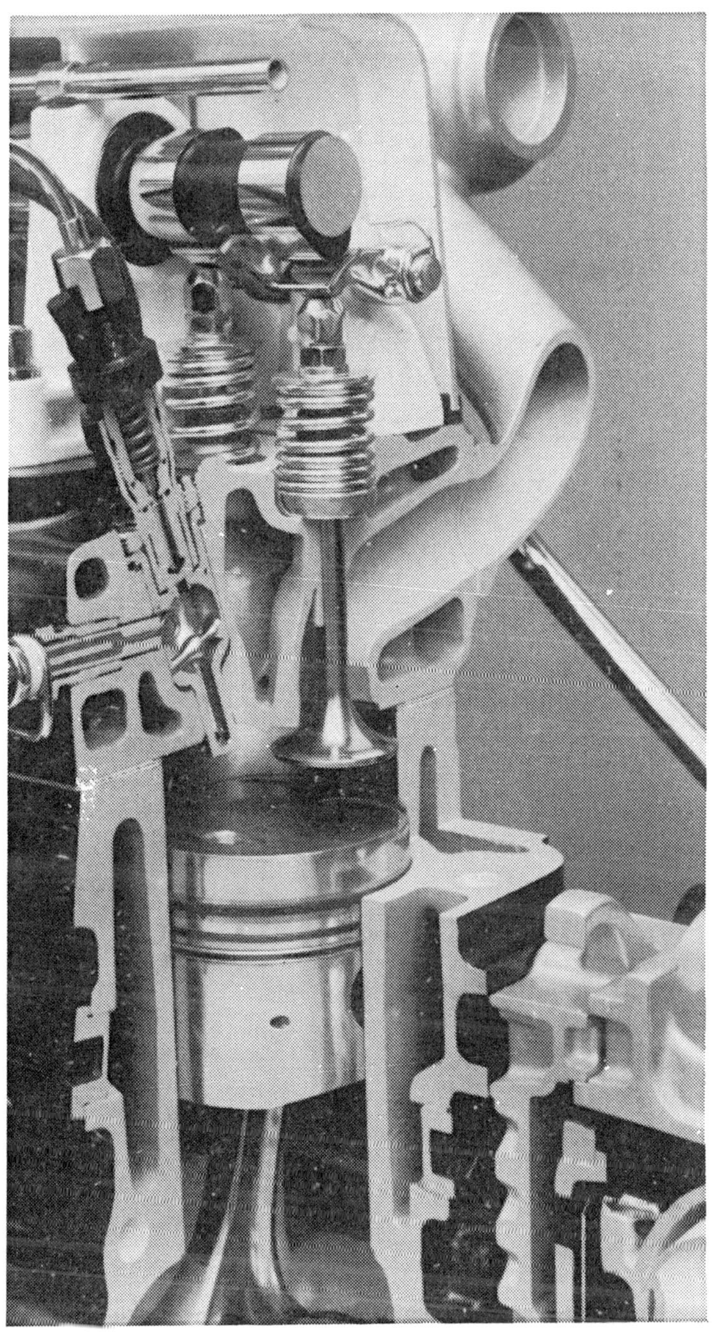

Fig. 1-3. The 300D in cutaway. Nearly all parts are common to the 240D.

Careful detailing has eliminated most of the vibration one could expect from the five-cylinder configuration. Crankshaft counterweights, working in concert with hydraulic motor mounts, absorb most of the unbalanced forces; the engine is balanced after assembly; and the clutch on manual-transmission models sold in Europe incorporates a sophisticated dampening mechanism. These features make the 300D smoother than the 240D.

Recently a 300D-powered Rekordwagen set three new diesel speed records by averaging 156.92 mph for 5000 miles. Its engine was essentially stock that had been fitted with Air-Research turbochargers for a 10-12 psi boost.

Mercedes-Benz is the most experienced builder of diesel passenger cars. Acceptance of their vehicles in the United States forms an interesting pattern. In 1958 M-B sold 808 diesel cars here, 10.9 percent of the firm's total U.S. sales. In 1965 diesel cars accounted for 28.2 percent of that total, but then interest in the diesel declined. And their sales fell to less than 15 percent of the total until the OPEC fuel crisis in 1973 hiked the price of crude oil. In 1975, 46 percent of American Mercedes-Benz buyers chose the diesel engine.

Peugeot follows Mercedes' philosophy to the extent of mounting gasoline or diesel engines on the same chassis. Because the diesel produces less power than the gasoline engine it replaces, the suspension members, steering components, and drive train elements are understressed and should give excellent service. When Peugeot marketed the diesel version of the 403 in this country, it had little success, but in 1974 the 504 was introduced. Offered in sedan and station wagon bodies, the diesel version displaced 2112 cc and developed 65 hp (DIN) (Fig. 1-4). With clutch and transmission, the four-cylinder, overhead-valve engine weighed 480 lb. Unlike most diesels, it had an aluminum head and a 22.2 to 1 compression ratio, perhaps as a result of aluminum's propensity to dissipate heat. The popularity of the 504 encouraged Peugeot to enlarge the engine by 141 cc for 1977. This overbored engine develops 71 hp (SAE net) at 4500 rpm, giving the sedan a slight performance edge over the Mercedes 240D.

The Volkswagen Rabbit (Fig. 1-5) may be the dawning of the age of the low-priced diesel. The engine (Fig. 1-6) is built along the same lines as the all-aluminum gasoline engine, sharing its major

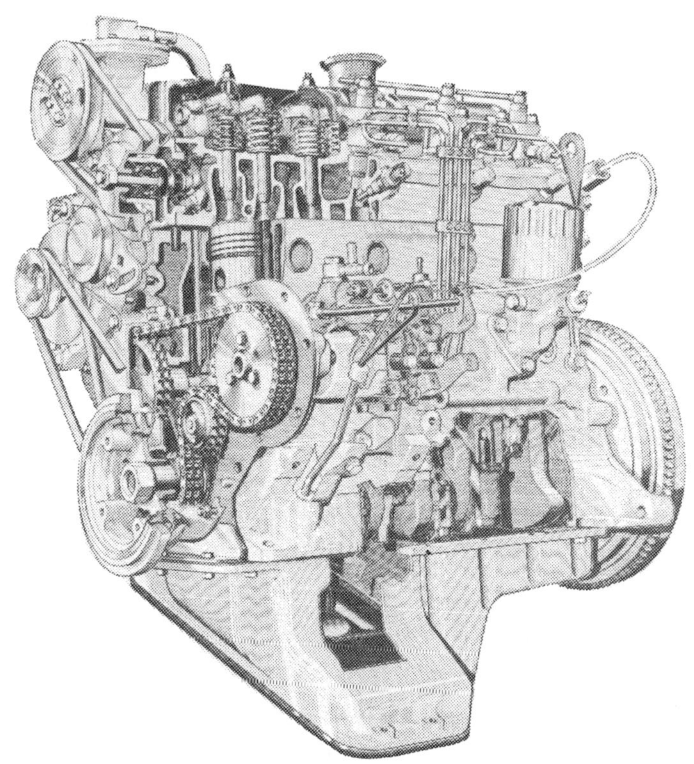

Fig. 1-4. The Peugeot XD 90 engine.

dimensions and components. The diesel block had originally been a special casting with additional oil passages and a beefed-up firedeck, but by current practice this block is used on the gasoline engine as well. The cylinder heads are quite different but share the same valves and belt-driven overhead cam mechanism, and the connecting rods and crankshaft are identical. In all, the diesel engine weighs 38 lb more than its sister. The Rabbit-D displaces 1471 cc and develops 41 hp at 5000 rpm, which is a terrific speed for a diesel engine. Since the diesel was finalized, the *gasoline* version has been bored out 1588 cc and develops 78 hp.

A particularly noteworthy feature of the Rabbit-D is the manual advance (Fig. 1-7) for starting, which eliminates much of the smoke generated by a cold engine and some of the clatter. It is new ground for diesel engines when service intervals are the same for both diesel

Fig. 1-5. If a diesel car could be called sprightly, this is the one.

and gasoline versions: oil changes are recommended at 7500 miles and filter changes at 15,000.

The Rabbit is the best performing, most agile, diesel car that money can buy. It has the best interior space utilization and is supported by the most extensive dealer network. Volkswagen has tested the diesel on the bench and in the field. Three hundred test cars totalled up ten million miles before the engineers were satisfied.

The question of durability remains. How long will the Rabbit run? By the accumulated experience of most diesel engineers, it won't stay together long. Even with more lightly stressed gasoline engines, Volkswagen Werke has had quality control problems: snapped exhaust valves, stripped cylinder-barrel studs, and soft

Fig. 1-6 The Volkswagen engine represents a revolution in diesel thinking.

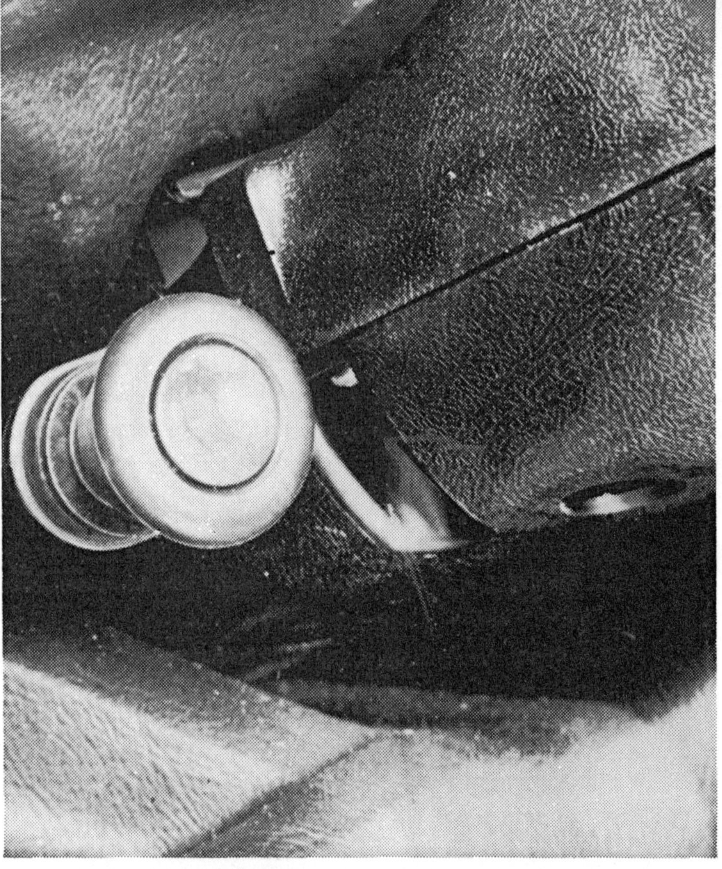

Fig. 1-7. Manual advance cleans and quiets the warmup.

crankshafts (on early production Rabbits). Yet, in engineering as in war, audacity is sometimes rewarded. Before the turn of the century De Dion Bouton was roundly castigated because of its high-speed engines. The critics were wrong, and De Dion Bouton became for a time the world's largest automobile manufacturer. Honda motorcycle engines suffered similar criticism: they were too light, turned too fast, and broke such sacrosanct rules as running wrist pins directly on steel connecting rods. In fact, Honda engines have set new standards for durability.

American manufacturers have offered diesel power for several years, but have created little publicity for their wares. Chrysler gives the option on certain off-road and recreational vehicles, and Ford has apparently built diesel sedans on special order.

From 1963 through 1968 International supplied customers with D-301 engines in pickup trucks and in the Travelall utility wagon. Then, in 1976, International announced plans to release 400 Scouts with Chrysler-Nissan six-cylinder diesel engines. These IHC vehicles can approach 80 mph, have tolerable acceleration, and consume as little as a gallon of fuel oil in 28 miles of steady-speed operation. Response to the $9000 vehicle was more than anticipated. Production climbed to 1200, then 2000, with dealers clamoring for more. During the year, the option was extended to Traveler wagons and Terra pickup trucks.

International settled on the Chrysler-Nissan CN6-33 after careful deliberation. Three other engines had been evaluated—Perkins, Peugeot, and one of the IHC small-block sixes. The IHC was too large; the Perkins fell down because of poor fuel economy and the problematic availability of parts; and the Peugoet, although the most economical of the group, presented installation difficulties. Because Chrysler distributes the Nissan in this hemisphere, spare parts were no problem. The engine also bolts up to Torque-Flite and other American transmissions.

Of course there were minor difficulties with this conversion as with any other. The automatic transmission shift points had to be changed to match the engine's torque characteristics, assembly-line workers needed time to become familiar with metric fasteners, and, embarrassingly enough, a production bottleneck developed because no one had thought to provide diesel fuel so that the vehicles could drive triumphantly from the end of the production line.

Other manufacturers are jumping on the bandwagon and entering the diesel-car market. Oldsmobile has tooled for a dieselized version of its 350 cubic inch V-8; Alfa-Romeo sells diesel cars in Europe and is expected to import them here; and Opel is building 25,000 diesel cars a year at Russelsheim. Whether or not these cars will be sold in this country is a moot point: GM admits to using the Opel project as a way of gathering experience with small compression-ignition engines.

Federal & State government agencies are also taking a closer look at the diesel.

The U.S. Urban Mass Transportation Administration has funded a diesel-vs.-gasoline test for New York City taxicabs. After

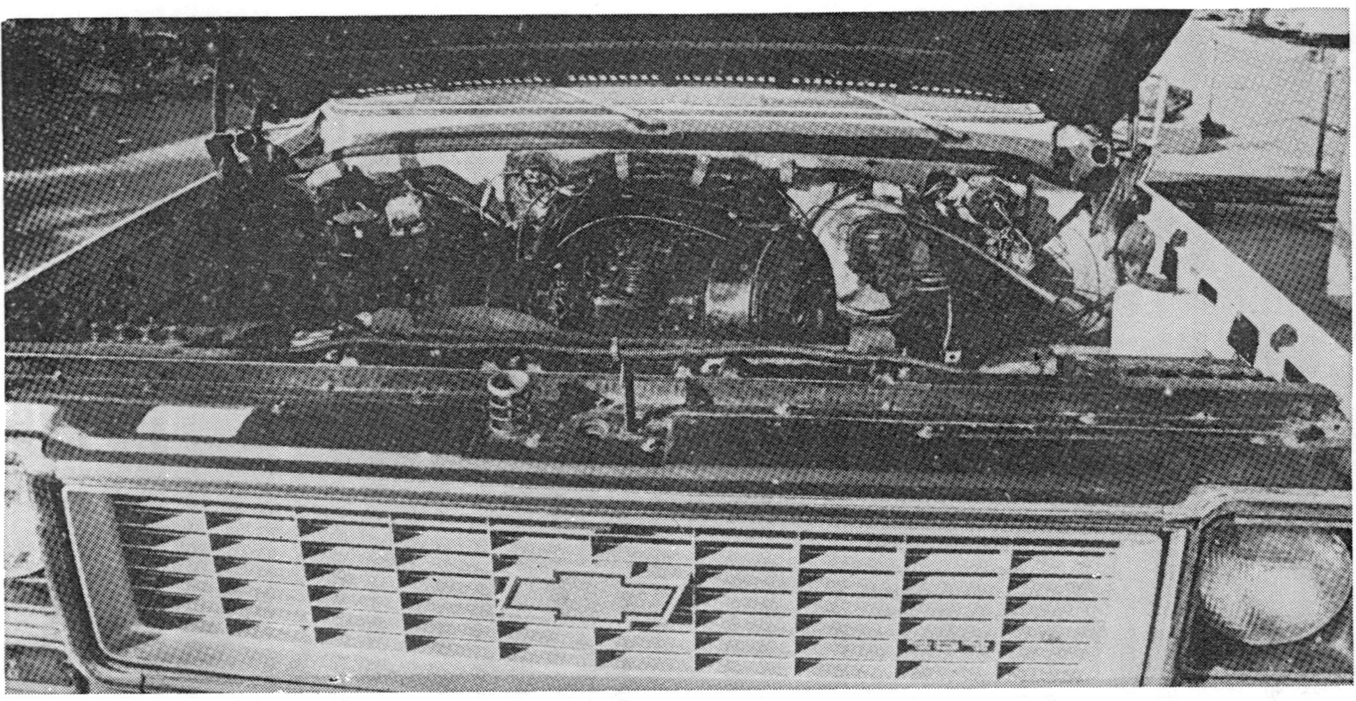

Fig. 1-8. A Hanomag engine boosted Chevrolet's milage from 5 to 15 mpg. (Courtesy RV World Magazine.)

the six-month delay that was spent unravelling red tape and prying an exemption loose from the EPA, 62 Dodge taxis were repowered with CN6-33 Chrysler-Nissan engines and another 62 were left untouched to serve as the control group. Dr. Richard Thaler, who is in charge of the project, reports that the small-scale tests show the diesel cabs are averaging 18-19 mpg in city traffic. The "Slant-Six" versions are burning twice as much fuel (8.5-9.5 mpg) and must be taken out of service twice a day to replenish their tanks. The cost of conversion was about $3000 per car, an expense that is justified in the eyes of many cabbies who "hunger" for fuel savings, to quote Dr. Thaler.

The California Department of Transportation (Caltrans) is also sponsoring vehicle evaluation studies. In 1973 two three-quarter ton Dodge pickups were repowered with Chrysler-Nissan 6-33 diesels. One engine was installed with a turbocharger while the other was turbocharged after 10,000 miles. Both engines gave 100 percent better fuel economy than the 360 cubic inch gasoline engines and had much lower emissions. Because Caltrans purchases through competitive bidding, the department was able to buy the naturally aspirated CN6-33 for $1640 and the turbocharged version for $360 more.

In view of the increased recognition of the diesel engine's value, it is hardly surprising that a minor industry has grown up to convert gasoline-powered cars and trucks to diesel (Fig. 1-8).

C-V Diesels in State Line, Maryland, (Fig. 1-9) made a number of conversions using Chrysler-Nissan and Mercedes-Benz engines. John Hostetter, owner and head wrench, knows diesels from theory and from years of hands-on experience. If you're ever on the Mason-Dixon Line, you will be welcomed at C-V Diesels. John and his crew are always busy, but never too hurried to stop and have a cup of coffee with you. Scattered about corners of the shop you'll find all sorts of strange hardware, such as an IHC V-8 gas engine with a turbocharger and on-demand water injection or a Hercules diesel that John plans to put in an automobile once he has its weight down and its power up.

Some of the most spectacular conversions made have been the work of Tony Capanna of the Wilcap Company in Torrance, California. One of these conversions, a Chrysler-Nissan powered Valiant,

Fig. 1-9. C-V Diesel is a well known conversion house in State Line, Maryland.

won the Internal Combustion Reduced Emissions Rally. This annual event, sponsored jointly by Cal Tech, Northrop Institute and UC Davis, is a showplace for some exotic machinery, but nothing there could match the CN6-33's near-zero emissions and 43 mpg. Another Capanna conversion, a Pinto with a CN4-33, achieved 60 mpg under ideal conditions. Like most converters, Wilcap is willing to work with an individual, offering advice and machine-shop services as needed. An example is the Mitsubishi-powered Dodge pickup (Fig. 1-10) built by Ray B. Smith with the support of the Wilcap crew. The transmission, a New Process 455 with close-ratio gearing, is driven through a Dualmatic overdrive. Unlike the more familiar Borg-Warner and Leycock planetary overdrives, the Dualmatic employs paired gears on a countershaft to give a 27% overdrive and can be engaged in any gear. Cruise control has been fitted for further economy.

In another case, a Rajay Model 301 F 70 turbocharger provides a maximum of 10 lb boost. The engine's connecting rods were beefed-up to withstand the greater combustion pressures with a handmade five-tube radiator to handle the thermal loads. Convenience modifications include air-conditioning, a 72-gallon reserve tank, a trailer hitch, and a camper shell. Weighing in at 5400 lb, this vehicle has adequate acceleration and gives 29 mpg at a steady 55 mph.

In Chicago, S&S Equipment Sales, Inc., was one of the first shops to specialize in repowering. Its customers are primarily commercial users who want the dependability and fuel economy of diesel power in light equipment. An S&S-installed Chrysler-Nissan increased the productivity of a Vermeer trenching machine by 250% and cut its fuel costs by 60% according to the machine's owner. The firm's first automotive conversion traded a CN6-33 for a tired Chevrolet 350 in a delivery van. The van was no longer daunted by Chicago hills and gave a consistent 20 mpg.

A

B

Fig. 1-10. Ray Smith's long-distance hauler (view A) with Mitsubishi power (view B). (Courtesy Diesel Car Digest.)

S&S has cut costs by purchasing new trucks from local dealers and returning the original engine for credit. Adapter plates are in stock to mate the Chrysler-Nissan engines with any Borg-Warner or New Process Gear transmission. Firewall, suspension, fuel tankage, and instrumentation modifications are kept at a minimum by the careful selection of the vehicle. Since commercial trucks have plenty of battery power, no additional batteries are necessary. Only a glow-plug indicator lamp is added to the instrument array. The average turn-around time is five days.

I do not want to give the impression that these three outfits are the only ones specializing in or capable of diesel conversions. They are typical. You will find others like them: small shops that will work with their customers to give personalized service; fairly large establishments with an immense array of technological know-how; and firms that have made diesel swapping almost an assembly-line operation.

A mechanically-adept owner can install his own diesel engine. Many industrial and truck diesels have bell housings which are compatible with standard automobile transmissions, making this part of the job a bolt-up. Where this is not possible, an adapter plate can be fabricated.

Most of these conversions employ used engines that have been purchased from wrecking yards. The going price for a Mercedes-Benz engine is $150-200 on the east coast.

Buying an as-is engine is always chancy—you might get it running for less than $1000 worth of parts and pump service, if you're lucky. Assuming transmission compatibility, your installation costs should be almost nil.

COSTS

Dr. Diesel readily admitted that his engine was expensive and predicted that it would always be. Its high working pressures demand high quality materials and a level of precision unknown in spark-ignition engines. The injector and fuel-pump parts are lapped to 0.0004 inch and finished with such perfection that a fingerprint is fatal. Diesel oil, as delivered from the refinery, is abrasive to these parts and must be passed through at least two stages of filtration before it may safely enter the pump. Water contamination is such a serious matter that some manufacturers suggest the engine not be refueled in the rain. A few vagrant drops of water mixed with a tankful of oil are enough to damage the pump. Indeed, mechanical precision makes such demands upon fuel purity that some observers doubt that the oil companies could supply enough uncontaminated fuel if a sizable fraction of American passenger cars were converted to diesel.

It is not then surprising that a Bosch inline pump lists for $750 or that the Chrysler-Nissan CN6-33 costs $3400. As a factory-installed option in the IHC Scout II, Terra, and Traveler, this 92 hp engine adds about $2000 to the list price. Nor is it surprising that the Mercedes-Benz 300D, the most sophisticated diesel passenger car built, has a price tag of approximately $17,000 a copy.

DURABILITY

While there is nothing inherent in the original design of the diesel engine to make it durable, most of them are. Engineers have deliberately over-designed them so that the customer can recoup some of their high initial cost. For example, Detroit Diesel bus engines are under warranty for 100,000 miles, GM is understanding for the first quarter of a million, and long-distance trucks travel three-quarters of a million miles between overhauls. Mercedes-Benz does not have to bandy about figures: the longevity of their automobile engines is legendary. One of the first M-B diesels, a 260 series built in 1936, ran for 13 years and 808,000 miles before it was retired. Warren Angstadt of Wernersville, Penn., has driven his 1953 170D 789,000 miles, with an engine change at 625,000 miles. Although Peugeot and VW Rabbits are too new to have accumulated this kind of milage, Peugeot cautiously estimates that their engines are good for 150,000 miles between overhauls, while VW claims that its diesel will have twice the life of its spark-ignition sibling.

POWER

The $17,000 Mercedes Benz 300D mentioned earlier weighs 3550 lb and develops 77 hp at the flywheel for a weight-to-power ratio of 46 lb per horsepower. The 240D, the four-cylinder version of the same car, totes 53.7 lb per horsepower. The 240D must be accelerated at full throttle for 23 seconds to reach the 55 mph speed limit. Woe betide you if you should encounter any hills. Other diesels, such as Peugeot, IHC, Checker, and Fiat, have the same clunky performance. The VW Rabbit is the fastest of the lot and accelerates through the quarter mile in just over 20 seconds.

Diesel engines have always been large and heavy for the power they deliver. The Detroit Diesel, which may be the most admired of all and is certainly the most original, checks in at 6.5 lb/hp in the 12V-71 version. The Cummings NT-380 weighs 7.6 lb/hp with turbocharger. In contrast, a well set-up gasoline engine, without forced aspiration, should weigh no more than 1 lb/hp. The expensive, high-technology fuel system and the fact that diesel-bearing pressures are more exigent than spark-fired engines (whereby diesel crankpins suffer almost as much at idle as at high rpm) presuppose a large, heavy engine. Dr. Diesel and his successors concentrated upon the stationary and marine markets, selling their products with the promise that their long-term durability would recoup the initial cost. It would have been folly to have compromised engine life by scrimping on a few pounds of cast iron or forcing up the rev limit for more output.

MAINTENANCE

Not having ignition points or carburetor, a diesel engine does not require periodic tuneups. Also, because its exhaust is less corrosive than a gasoline engine's, the exhaust system can be expected to last the lifetime of the car. The engine accessories are usually better designed than for gasoline engines and should last longer.

On the other hand, high compression ratios, which are on the order of 20 to 1, mean blowby around the piston rings. The lubricating oil becomes contaminated by fuel and must be changed at frequent intervals. Some manufacturers suggest the change be made every 2000 miles.

Although major breakdowns are rare, they are costly. Because of limited production runs and high standards of quality, diesel parts are expensive. Piston-and-ring sets for the Mercedes-Benz 220D cost approximately $400. Pistons for the earlier 180D retail at $52.40 each, less rings and wrist pin.

EMISSIONS

In spite of the visible smoke on cold starts and during acceleration, diesel engines are the cleanest internal combustion engines known. Diesels operate with a surplus of air that eliminates most hydrocarbon and carbon monoxide emissions. Because of lower flame temperatures, oxides of nitrogen (NO_x) are only a small fraction of what they are for gasoline engines.

The only emission control on current diesel engines has been provision to recycle crankcase vapors to the combustion chambers. The plumbing is simple and, if it operates correctly, has no effect upon engine efficiency. In contrast, modern gasoline engines have an array of emission-controls so complex that no general mechanic could service all of them.

But as of 1978, with new standards, the honeymoon is over. Formerly we were on the 1975 standards mandated bythe Clean Air Act that allowed 2.0 grams of NO_x per mile. The diesel comfortably meets this while puffing out 1.5 grams per mile. The tightened 1978 standards limit NO_x emissions to 0.4 gram per mile and may effectively end the diesel as a commercial automotive power plant. No one has yet discovered how to tame the diesel combustion processes. Exhaust gas recirculation, the classic method of controling NO_x in gasoline engines, does not work for diesels.

These federal standards apply to NEW vehicles. Once sold, the owner can install any engine he wishes, *so long as it meets applicable state emissions standards*.

DRIVEABILITY

Driving a diesel automobile is not much different from driving one that is powered by gasoline. The acceleration is poor, but the flat torque curves of most compression-ignition engines gives a sense of responsiveness. Diesel-idle may be lumpy, depending upon the engine. All engines are mechanically noisy; but, once underway, this noise tends to be muffled.

Cold starts can be an embarrassment, although the situation has improved in the past few years. The drill is to switch on the heater plugs, which are resistance elements not unlike glow plugs in model airplane engines, and wait for 30 seconds or so until the chambers are warmed. In extremely cold weather, you may follow the advice of Ettore Bugatti to one of his customers—invest in a heated garage! As a less expensive solution, you may find an immersion-type oil heater to be helpful. Starting fluid should not be used: it will wash the oil off the bores and detonate.

Once the engine does come to life, you can expect it to smoke and detonate for a few minutes.

FUEL ECONOMY

The biggest advantage of the diesel is its superior fuel economy. Examples of this have already been given in the text, but the point can bear amplification.

The thermal efficiency of an engine is the percentage of heat extracted from the fuel and converted into useful work. An engine that converted all the energy that was latent in the fuel into work-energy to turn its crankshaft would have a thermal efficiency of 100 percent. So far human ingenuity has not succeeded in breaking the Third Law of Thermodynamics by inventing such a 100% thermally efficient engine that would lose no heat to the fuel system, none to internal friction, and none to the exhaust. We have to settle for something less.

Under the best conditions a gasoline engine can achieve thermal efficiencies on the order of 35%. A diesel engine can do somewhat better and may achieve 40%. In practice, thermal efficiencies for both engines are less, but the diesel does consistently remain ahead of its gasoline counterpart. The figures of Table 1-1, supplied by Mercedes-Benz, are not unreasonable.

Table 1-1.

Operating Mode	Thermal Efficiency	
	Gasoline passenger-car engine	Diesel passenger-car engine (indirect injection)
Light load Low speed	13-20 %	20-25 %
Part load Medium speed	29 %	33 %
Full load Medium speed	26 %	30 %

Other economies arise from the different nature of the two fuels. While fuel blends vary, particularly in the United States, it is fair enough to say that diesel oil and gasoline have the same heat energy per *pound*. But diesel oil is denser than gasoline, and so has a greater heat content per gallon. Standard reference books credit diesel fuel with 130,300 BTU (British Thermal Units) per gallon and list gasoline as having 116,400 BTU/gal. Since fuel is purchased by the gallon and consumed by the pound, diesel oil would give some operating economy by virtue of its heat content alone.

In addition, diesel oil is cheaper to refine than gasoline because it requires less energy and less capital expenditure. This savings is reflected in the price. In most areas diesel oil is also less heavily taxed, thanks to the activity of the trucking lobby. Thus you can drive a diesel car farther on a gallon of fuel for which you will have paid less. Figure 1-11 compares fuel costs for equivalent diesel and gasoline engines over a 2000-hour operating period.

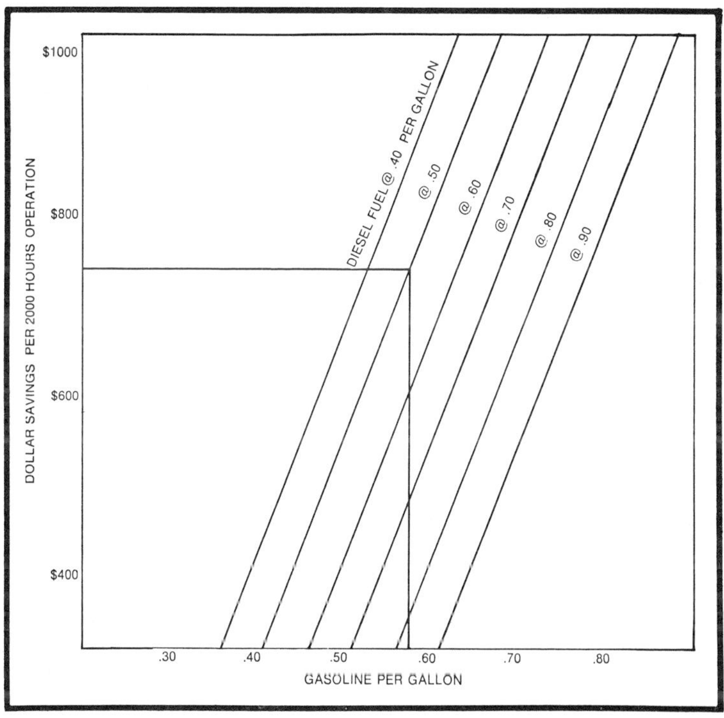

Fig. 1-11. Dollar savings with diesel power. (Courtesy Onan.)

While no one welcomes the prospect, there will come a day when petroleum supplies fall short of demand. Prices will rise sharply and the government can be expected to divert as much petroleum as possible away from private transportation and into more vital areas. Diesel fuel would be among the last of the petroleum products to disappear, since it is essentially the same as No. 2 heating oil. Some converters are already risking the wrath of the Treasury Department by using heating oil in their vehicles. The government fumes about this being a tax evasion.

SUMMARY

In conclusion, diesel cars are practical alternatives to gasoline-powered cars if the owner is willing to accept some loss of performance. Diesel engines are expensive, but the higher initial cost is repaid by operating economies. On the whole, diesel engines last longer than gasoline engines, have slightly lower maintenance costs, and much lower fuel costs.

And it is practical to convert a gasoline-engined car or truck to diesel, using a new or reconditioned engine. The cost will vary depending upon the installation. A simple conversion should cost about $3500, using a new Chrysler or equivalent engine. Purchasing a used engine, rebuilding it, and installing it yourself can keep costs well under $1000.

The rest of this book will tell you how to go about such a conversion.

Diesel Basics

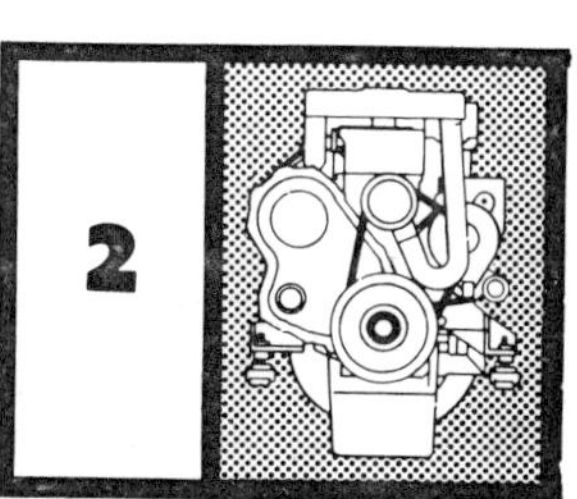

Diesel engines are, in broad outline, similar to gasoline engines. Pistons, rings, valves, crankshaft, and flywheel have the same basic function and are called by the same names (Fig. 2-1). However, because diesel compression ratios are much higher, mechanical parts for them are correspondingly heftier. The differences between the engines become more apparent as we move toward the cylinder head. Instead of spark plugs and carburetor, a diesel engine has injectors and an injector pump. In most diesels, injectors are fed by a high pressure pump mounted on the side of the block and driven by the crankshaft. A few engines combine the injector and the pump into single units, one for each cylinder. The intake manifold, or header, admits air into the engine and may be topped by a turbocharger for additional power.

OPERATION

The operation for a diesel, or compression-ignition, engine differs from that of a gasoline, or spark-ignition, engine in three ways:

1. Air is compressed in the cylinder before fuel is introduced.
2. The amount of air is the same, regardless of the throttle setting.

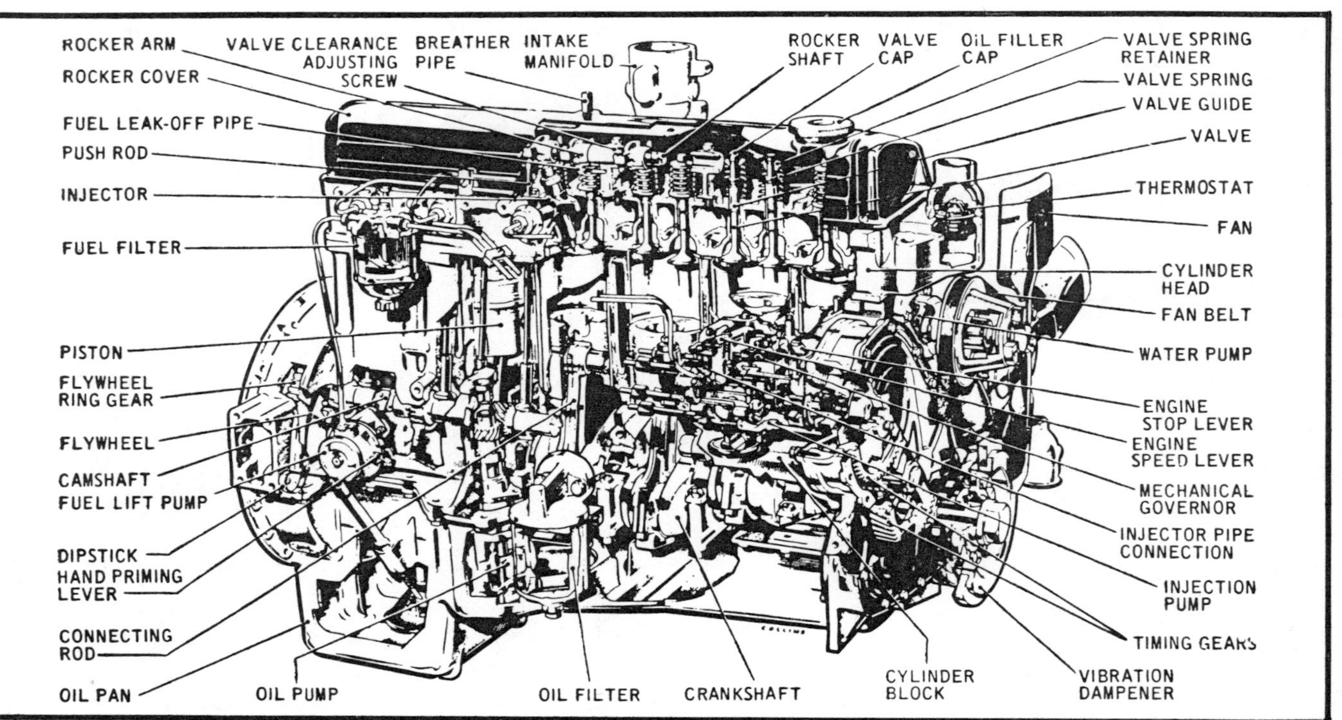

Fig. 2-1. Diesel nomenclature. (Courtesy Industrial Engines and Turbine Operations, Ford Motor Co.)

3. The fuel is ignited by the heat of compression. Each of these characteristics deserves comment.

When a gas such as air is compressed, the molecules collide with and rebound from each other, like self-energized pool-balls run amuck. Heat results from this molecular movement. In order to generate this heat, automotive diesel engines must have compression ratios as high as 23 to 1. At the bottom of the stroke the volume of the gases in the cylinder is 23 times as large as the volume of the gases at the top of the stroke. Turbocharging increases the effective compression ratio, since it packs more air into the cylinder.

Because the fuel is introduced late in the cycle, when the piston is approaching top dead center, it must be forced in against the compression pressure. It is currently the practice to inject the fuel at between 1300 and 2600 psi.

The injectors break the fuel into a spray for easy ignition. Because the temperature of the air in the cylinder is already 700° F or higher, the fuel begins to vaporize immediately. After a short ignition delay, which will be discussed later, the fuel ignites. Pressure build-up is rapid at first, and then slows as the piston continues on the downstroke. The synchronizing of the pump with the piston so that maximum pressure occurs just when the piston goes over top dead center is called "timing" the engine.

Engine speed is determined by the amount of fuel injected during each cycle. At idle the fuel may only account for 1/80,000 of the cylinder volume, a fraction that gives a good indication of the precision of diesel pumps and injectors. At full throttle the amount of fuel injected is about 1/20,000 of the cylinder volume. The amount of air remains constant, and always is more than is needed for combustion.

This brings us to a small problem encountered in conversions. The amount of air that enters a gasoline engine varies with speed and load. Vacuum is developed in the intake manifold according to the position of the throttle plate. At idle the throttle plate is almost closed and manifold vacuum can draw 18 – 22 inches of mercury. Vacuum decreases as the throttle valve is opened and the engine "breathes" easier. Manifold vacuum, the vacuum existing between the intake valve and the throttle plate, is used as an input by most automatic transmissions. A diesel engine has a fairly constant header (which is the part that corresponds to the manifold) vacuum, caused by the air-cleaner restriction. This vacuum is no indicator of demands on the engine and cannot be used to signal shifting.

Thus, if you want to use an automatic transmission with a diesel engine, you will have to use a TorqueFlite which does not depend upon vacuum, or else you will have to do some very inventive and fancy engineering.

OPERATING CYCLES

Figure 2-2 shows a Detroit Diesel going through its paces. This is a two-cycle engine. Indeed, it is the world's most successful and respected two-cycle engine, and is used in trucks, buses, small boats, and in a few engine swaps. As with all two-cycle engines, there is one power impulse at every revolution of the crankshaft. You can also look at this another way: there is one power impulse per two stroke-cycles of the piston. The term "stroke-cycle" will become clear later.

This particular engine uses a Rootes-type blower to "scavenge" (clear) cna charge the cylinder. Other two-cycle diesels use the back of the piston as a charging pump, following the model set by two-cycle gasoline engines.

In the case of the Detroit Diesel, as the piston falls to its lowest point for the scavenging phase it uncovers a ring of ports milled into the side of the liner. The exhaust valve is also open. The blower forces air through the ports, into the cylinder, and out over the open exhaust valve, clearing the chamber of spent gases. The exhaust valve closes; the piston moves upwards, compressing the air before it. Near top dead center (TDC) the injector sprays fuel into the chamber. The combustion pressure peaks as the piston goes over top dead center: this is the power stroke of the cycle. The piston is forced down, but this expansion stroke is cut short by the opening of the exhaust valve just before the piston again uncovers the air-inlet ports. The exhaust gases "blow down" until the pressure in the cylinder is equal to atmospheric pressure. The gases that remain are cleaned out as the scavenging phase begins.

Most automotive engines are built on the four-cycle principle (Fig. 2-3). A four-cycle engine provides a power impulse every other revolution of the crankshaft, at every fourth stroke of the piston.

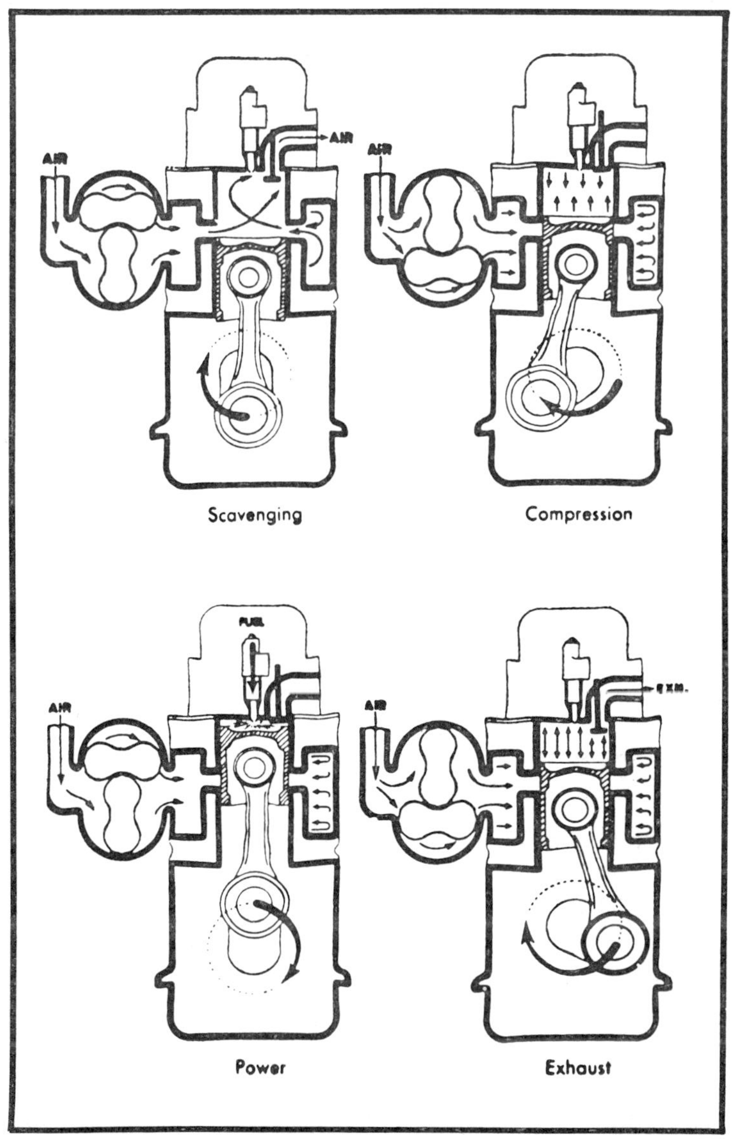

Fig. 2-2. Two-cycle operation. (Courtesy Detroit Diesel Allison.)

Each cycle—intake, compression, combustion, and exhaust—roughly corresponds to an up-stroke or a down-stroke of the piston.

The intake valve opens to admit air as the piston descends on the intake stroke. Once the piston passes bottom dead center (BDC), the intake valve closes and compression begins. The injector

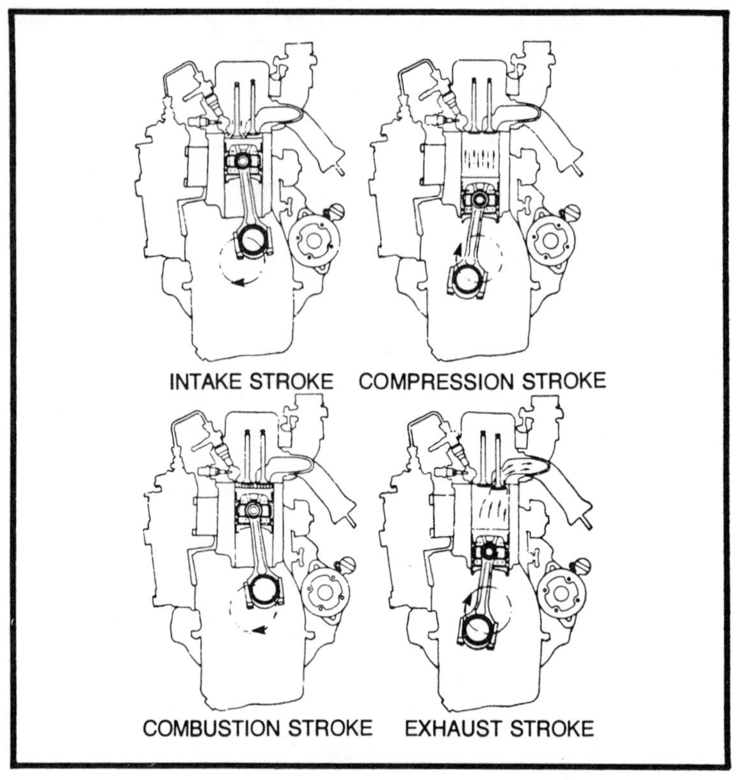

Fig. 2-3. Four-cycle operation. (Courtesy Chrysler Corp.)

releases a spray of fuel into the cylinder just before the piston reaches TDC. Burning oil and superheated air drive the piston down on the combustion stroke as it is labeled in the drawing. "Expansion" or "power" stroke are the more descriptive terms that are more commonly used. The exhaust valve opens before the piston reaches BDC. Those gases that do not blow down past the valve are purged from the cylinder as the piston rises on the exhaust stroke.

Two-cycle engines have the advantage so far as power, size, and weight are concerned. The Detroit Diesel 4-71 develops 35.48 horsepower/liter. In comparison, the Perkins 4.99, which is one of the most advanced four-cycles, manages a bit more than 26 hp/l. Other four-cycles fare worse: the early production Ford developed 16.5 hp/l; the Fiat, 21 hp/l. The high power output of the 4-71 gives at a rating of 4.8 lb per hp. The Perkins 4.99 is nearly twice as heavy at 9.05 lb/hp.

In part, this two-cycle performance is the happy result of one power pulse per crankshaft revolution. And, as far as the 4-71 is concerned, it is also partly due to the boost provided by the blower.

DETONATION

This is the most appropriate time to discuss an unpleasant subject. Diesel engines can detonate. Normal, controlled combustion can be supplanted by maverick explosions that can blast holes in the piston tops. Gasoline engines suffer the same malady, but the mechanism is different: gasoline detonation occurs *after* the spark, when the flame front has moved to the outer edges of the combustion chamber. The fuel-air mixture that remains, being subject to increasing pressure and heat, may, if conditions are right, go off with a bang. Diesel detonation occurs early, *before* normal ignition. The graph in Fig. 2-4 helps to explain the situation. Injection commences at point A, but ignition does not occur until point B. The time period A-B is known as ignition lag. If the engine is going to detonate, it will do so sometime during the lag. Suddenly, all the fuel that has been injected will go off and the piston will be trapped between this explosion that is trying to drive it backwards and the inertia of the rotating parts of the engine that should send it over TDC. All very destructive.

COMBUSTION CHAMBER TYPES

The simplest combustion chamber utilizes a depression in the piston as the site of combustion. Figure 2-5 illustrates one variation

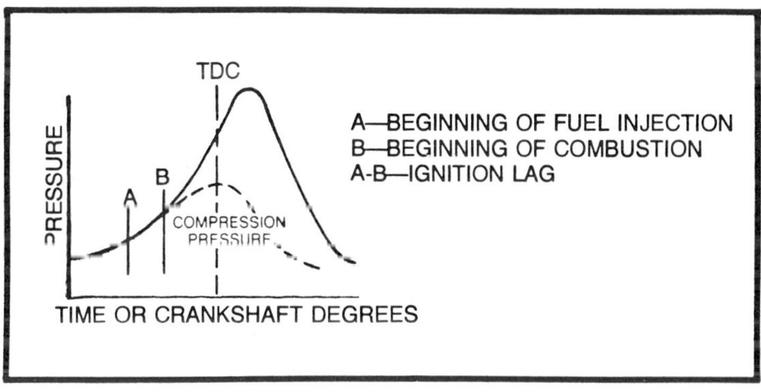

Fig. 2-4. An indicator diagram showing ignition lag.

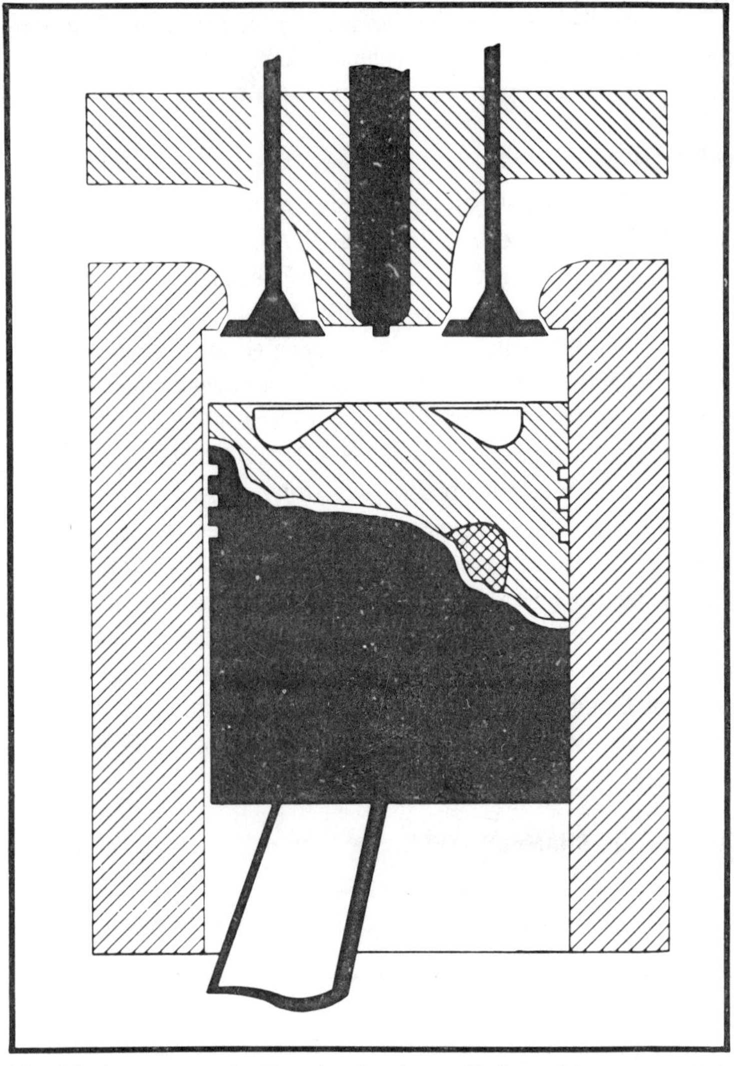

Fig. 2-5. An open combustion chamber is usually formed by one or more depressions cast in the piston top.

of this design, known as the "open," or "direct-injection" chamber. Mack chambers are similar to the one shown, except that the two piston depressions are joined at the center of the piston crown. Flats at the rim of the piston crown are "squish" areas. As the piston approaches top dead center, air is trapped between the flat edges of the piston and the roof of the chamber. As pressure increases, the air "squishes" inward, spilling over into the central depression, and speeding combustion.

This type of combustion chamber is highly efficient: because there is less chamber metal to heat, there is more heat available to do the work of turning the crankshaft. By the same token, cold starting is easier. These engines do not need glow plugs or other starting aids. Although the turbulence produced by the squish area is welcome, its amount of turbulence is hardly to be compared with that generated by a precombustion chamber. Less energy is used in the process.

There are two disadvantages to an open combustion chamber: it gives a harsh idle of the kind that you do not want in an automobile, and the exhaust emissions are higher than for indirect chambers.

The precombustion chamber was used on the earliest diesel engines and is still used for Caterpillar, Mercedes-Benz, and a number of smaller engines.

Figure 2-6 shows the basic arrangment: The precombustion chamber is connected to the main chamber by a narrow passage. Fuel is injected into the small chamber and where it begins to burn. As it burns, the pressure in the chamber increases, driving the partially burnt fuel charge out into the main chamber. The violence of this action assures complete combustion.

Since the fuel is dispersed by virtue of its own heat energy, the pump pressures can be reduced and the injector can have larger, coarser orifices. Less maintenance is required. Another benefit is very consistent fuel economy coupled with the ability to digest almost any liquid hydrocarbon. Mercedes engines that use this system can be run on kerosene or on a mixture of diesel oil and gasoline.

Precombustion chambers are not particularly efficient: the relatively large surface area absorbs heat, quenches the flame, and gulps fuel. These engines must employ high compression ratios, on the order of 20 to 1 and higher. While high compression ratios, which can be more exactly called high ratios of expansion, mean increased efficiency, they also mean a heavier hunk of an engine. The block and the working parts must be massive enough to withstand the force of compression, while the starter must be powerful enough to spin the flywheel.

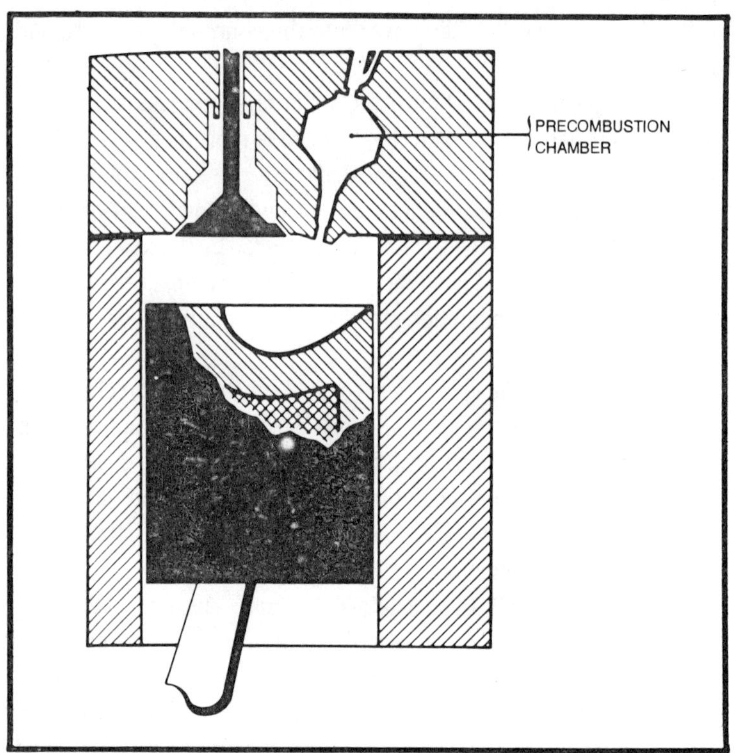

Fig. 2-6. A precombustion chamber is a small antechamber, connected to the main chamber by a passageway.

At first glance, a turbulence or "swirl" chamber (Fig. 2-7) looks like a precombustion chamber. There are however, several differences. The precombustion chamber is relatively small, accounting for 25-40% of the total chamber volume; its function is to begin the combustion process which is completed in the zone above the piston. The turbulence chamber is much larger, accounting for nearly all of the area above the piston. Figure 2-7 does not make this very clear, since the piston is not shown at TDC. If it were, the top of the piston would come to the lower edge of the passage connecting the two chambers. Although some combustion does occur above the piston, most of the action takes place in the turbulence chamber. The air and flame movement in to and out of the chamber has a definite swirl that gives the chamber its name.

The most widely used turbulence chambers are the Comet series, developed by Sir Harry Ricardo back in the 1930s. The

Comet Mark V is characterized by a small depression in the piston crown, a vertically mounted injector, and a tangential passage between the chamber and piston. Figure 2-8 illustrates the Peugeot version of the chamber; the Chrysler-Nissan and Volkswagen adaptations are similar.

The Comet V has two noteworthy advantages:

1. Less exhaust smoke, particularly at low speeds. Until this chamber was developed, the part-throttle output of diesel engines was limited by smoke.
2. Less noise across the rpm band, and particularly at idle.

On the other hand, the large surface area of the chamber makes glow plugs necessary.

There are other combustion chamber designs, but you are not likely to encounter any of them in engines suitable for automotive use.

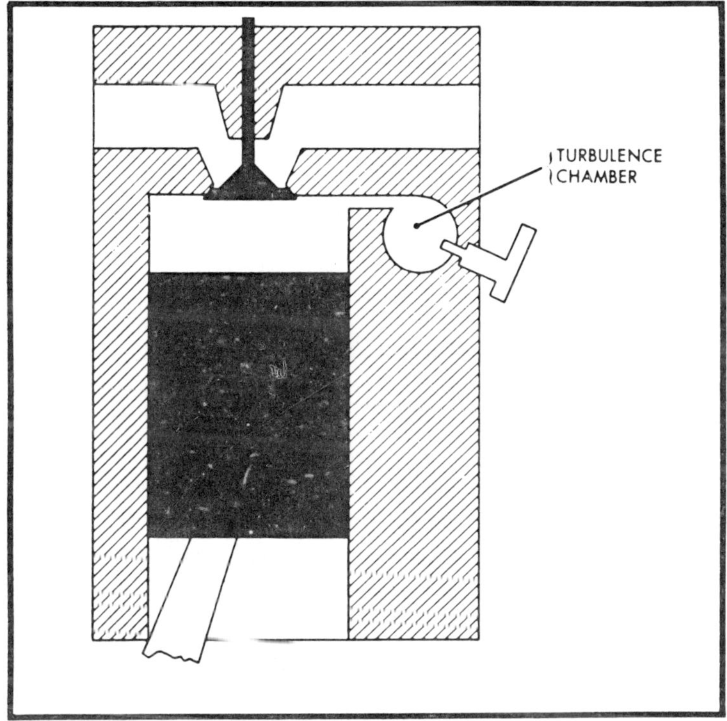

Fig. 2-7. A turbulence, or swirl, chamber occupies a large portion of the combustion space.

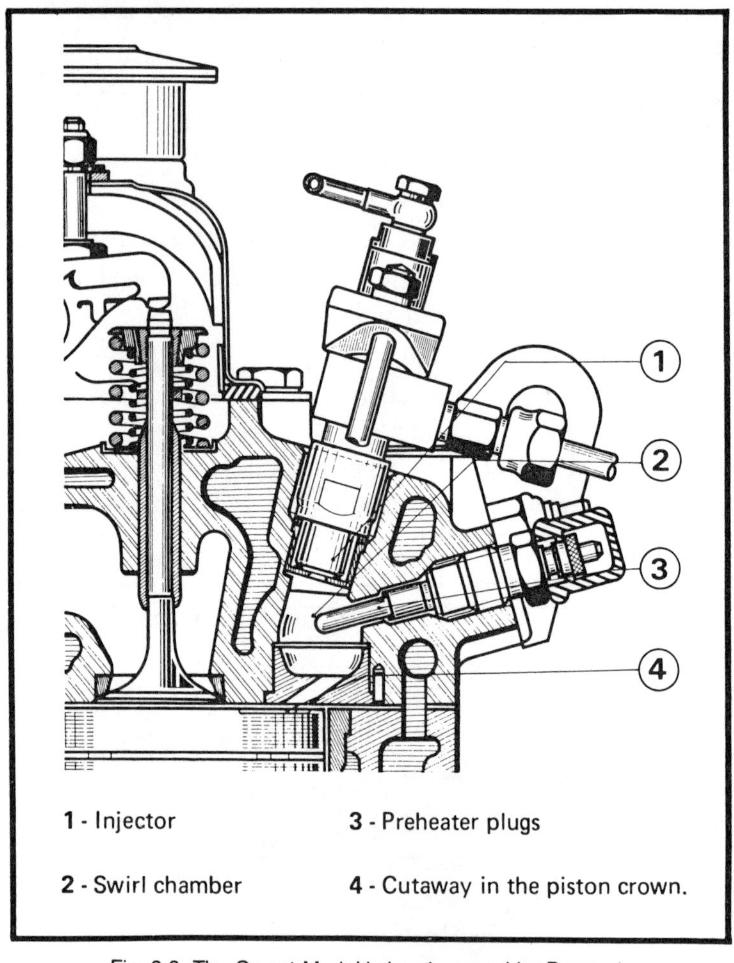

Fig. 2-8. The Comet Mark V chamber used by Peugeot.

TURBOCHARGING

A turbocharger is a centrifugal air pump, powered by the exhaust gases, and turning with fantastic speed in the neighborhood of 100,000 rpm (Fig. 2-9). Typical turbocharger applications are limited to 7 or 8 pounds per square inch boost, which means that the cylinders are pressurized to 7 or 8 psi above atmospheric pressure. More boost can be obtained, but greater pressure tends to be death on pistons and connecting rods.

Figure 2-10 shows the basic arrangement. The turbine wheel, which is the driving member, is made of special, heat-resistant alloy

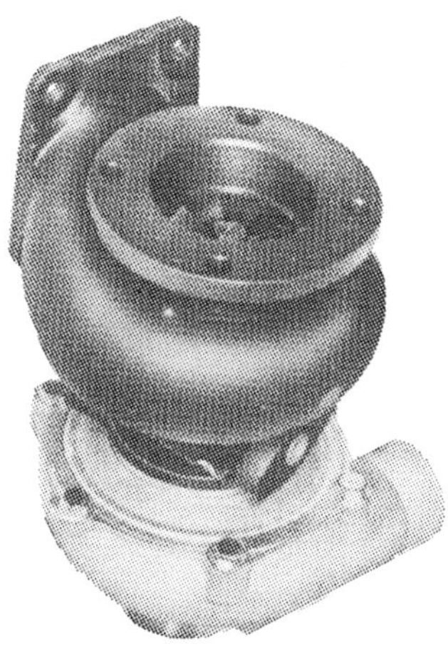

Fig. 2-9. The Accel TurboSonic turbocharger is installed with a patented priority valve that limits boost pressure to 6–8 psi.

steel because it must withstand the temperature of the exhaust gases, which, however, are not as hot as they would be in a gasoline engine.

Very little back pressure is developed across the turbine. Most of the energy bled for the gases is in the form of heat and velocity. The turbine shares a common shaft with the impeller wheel, which is the pump element. Both parts rotate together and, because the

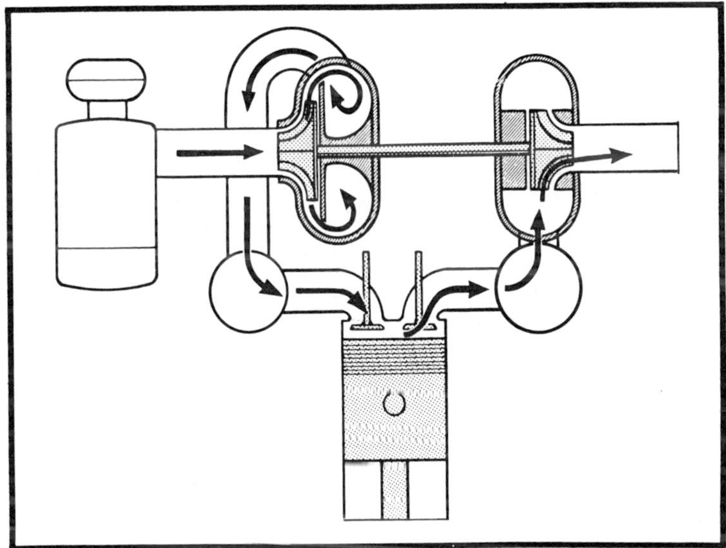

Fig 2-10. The basic turbocharger setup. (Courtesy Waukesha Motor Co.)

speed of rotation is so high, they must be in as perfect a balance as it is humanly possible to attain. The oiling circuit was omitted in the drawing: it consists of a feed line from the high pressure side of the engine lube oil pump and a return line to the crankcase. Oil flow to the shaft bearings must be positive, since air bubbles or any other interruption would mean the end of the turbocharger.

If in the above two paragraphs you have gotten the impression that turbocharging is strong medicine, your impression is correct. The power boost *is* remarkable. For example: the International Harvester 414 CID engine develops 157 hp in its naturally aspirated form, but, with a turbocharger, its output jumps to 220 hp and its fuel economy improves when figured on a horsepower-per-hour basis.

The increase in power output is made possible by the cooling effect of the additional air which allows the burning of more fuel without overheating the valves and pistons. The improvement in fuel economy comes from the better air distribution to the cylinders.

Turbocharging, when done by the factory, is expensive. Such installations can run up to $2000. In addition to the actual turbocharger, factory installations usually include heavier parts, additional oiling circuits, and other modifications to lower the compres-

Fig. 2-11. A turbocharged 200D built by George Baker. The valve limits vacuum to the governor.

sion ratio. Most diesel engines do not tolerate turbocharging without some detuning.

It is possible to add a turbocharger in the field. Figure 2-11 shows the installation on a Mercedes-Benz 200D. The engine was carefully rebuilt and the boost is limited. An unforeseen problem came from the venturi-type governor which is mounted in front of the turbocharger. At high speeds, the venturi was fooled by the additional airflow and signaled the pump to reduce fuel delivery. A valve in the venturi-to-diaphragm vacuum line solved this problem.

FUEL

Diesel engines will run on any liquid hydrocarbon, even Jim Bean or kerosene. However, the engines are happiest when they are fed their favorite density of liquid hydrocarbon-diesel oil.

Specifications. The American Society of Testing Materials (ASTM) has developed standards for diesel fuel oil. No. 1-D oil is recommended for normal use in automobiles, although some engine-makers suggest No. 2-D for trailer-towing and other extreme-duty applications.

Distillation value. The Fahrenheit temperature at which 90% of the fuel boils off is given as the figure for distillation value. Because smaller engines are not as tolerant of volatile fuels as are larger ones, No. 1-D has a maximum distillation temperature of 550°F. The distillation point for the No. 2-D fuel, intended for large engines, is 640°F.

Flash Point. The flash point is the lowest temperature at which the oil will release flammable vapors. Although meaningful to chemists, it has no direct bearing on engine operation.

Cetane Number. The cetane number rating is analogous to the octane number rating of gasoline. Both gasoline and diesel oil are distillates obtained from cracking crude petroleum, as are the other liquid hydrocarbons. As the crude is slowly heated, the shorter and lighter hydrocarbon chains are the first to be vaporized. They are captured and condensed, and the heat is turned up under the remaining brew to vaporize the hydrocarbons that are just a little longer and a little heavier. The process continues until only sludge is left in the vat. In cracking, gasoline and kerosene are among the earlier distillates. Diesel oil is heavier, and is distilled not long before the lube oils. "Octane" is the chemist's term for the preferred molecule of gasoline: a short, light, hydrocarbon chain with 8 carbon atoms and hydrogen atoms. "Cetane" is another chemist's term, this time for the preferred molecule of diesel fuel: a longer, heavier, hydrocarbon chain with 16 carbon atoms and 34 hydrogen atoms. A "high octane" gasoline has mostly 8-carbon hydrocarbons, with a few stray Siamese-twins that are 16-carbons, and other aberrations and mutations in the hydrocarbon family. A "high cetane" fuel has mostly 16-carbon hydrocarbons, with a few divorcees wandering around as 8-carbon hydrocarbons. You can see that the high octane gasoline would have a very low cetane count and be a very poor diesel fuel, and that the high cetane diesel oil would have a very low octane count and be a very poor substitute for gasoline.

The long-chain molecules of cetane are entangled and intertwined with each other like excelsior kindling. Pure cetane gives a very rapid ignition and has been assigned a cetane number of 100. (Logical? It *is* 100% cetane.) Alpha-methyl-naphthalene, which ignites in about as long as it takes to say its name, contains no cetane and has been given a cetane number of 0. These two hydrocarbons are mixed, and then matched against a sample of the fuel to be rated. A standard test engine is used in which the injector is timed for a piston position of 13 degrees before TDC. The compression ratio is then varied until ignition does occur at TDC. The cetane number of the fuel is set as being the percentage of cetane in the cetane/alpha-methyl-naphthalene mixture that required the same compression ratio.

Fuels with cetane numbers between 40 and 55 are desirable. Fuels rated lower than 35 or 30 are piston-breakers.

Viscosity. Viscosity is a measure of the "pourability" of the fuel, of its willingness to leave the tank and pass through the line into the high pressure pump. The viscosity is an important fuel characterestic, particularly in subzero temperatures. The thicker and more viscous the fuel, the less likely it is to freeze.

The ASTM specification is in centistokes (cST), the preferred way of expressing viscosity. The higher the number, the thicker the oil. Some refiners still use the old system that expresses viscosity as Saybolt Universal Seconds (SUS or SSU), so it is convenient to be able to convert cST into SUS. As a rough rule of thumb, cST × 5.5 = SUS.

IMPURITIES

Water and sediment are the major causes of injector and pump failure. One engine maker seriously recommends that engines not be refueled in the rain. Sulphur cakes-out in the combustion chamber, leaving harmful deposits. Ash, the solids left after the fuel burns, scores cylinder walls and abrades rings.

Can No. 2 heating oil be used in diesel engines? Most manufacturers agree with the federal government and say "No." While the

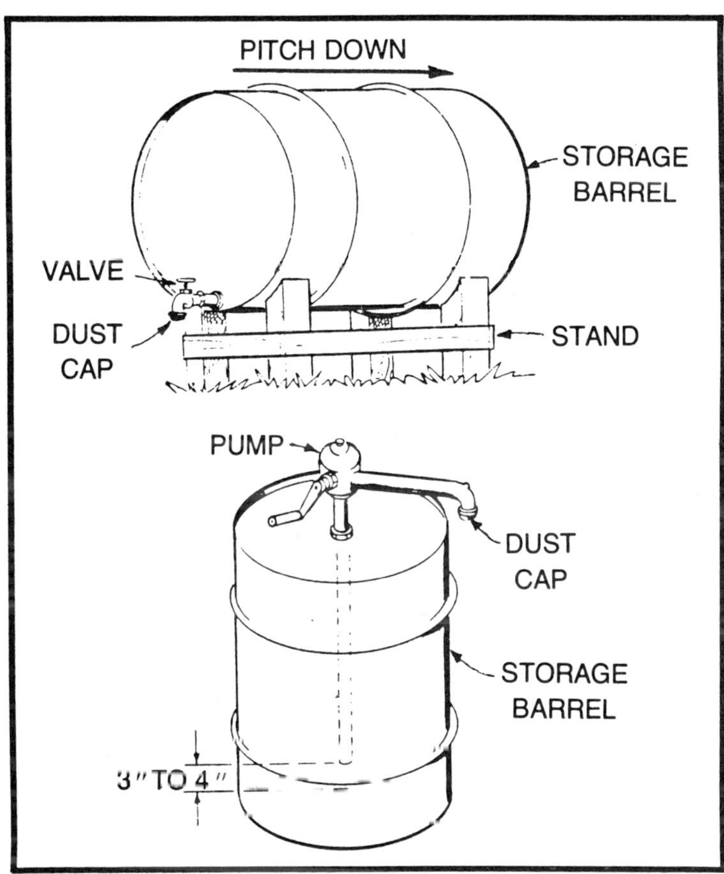

Fig. 2-12. Fuel oil drums should be mounted so that the dregs remain in the barrel. (Courtesy Onan.)

government objects to losing tax dollars, the manufacturers worry about the chamber deposits that might result from the use of furnace oil. Most noncommercial users take their chances with the chamber deposits and the feds. At this writing, diesel fuel costs about 55 cents a gallon on the east coast, and a few pennies less in the far west and deep south. Heating oil averages about 47 cents.

Fuel Storage and Handling

Most fuel contamination occurs during storage and particularly during the transfer of fuel from the holding tank to the vehicle. Observe these precautions.

- Do not store fuel in zinc or galvanized (terne) tanks. Zinc and fuel oil react to form harmful chemical compounds.
- Pitch holding tanks down from the outlet; if a vertical pump is used, adjust the length of the pump pickup so that it is above the bottom of the tank (Fig. 2-12).
- It is good practice to fit a filter and water trap to the tank outlet.
- Keep vehicle tanks as full as possible to reduce water condensation.

While diesel oil is difficult to ignite in the liquid state, the vapors are highly explosive. Refueling operations should be carried on away from any open flames. Obey the no smoking rule.

Swapping Engines 3

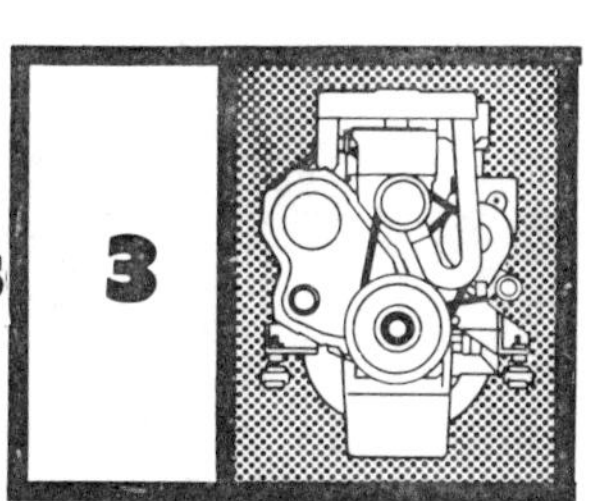

Installing a diesel engine in an automobile is an ambitious undertaking. It will require intelligent planning, attention to detail, and the determination to see the job through to completion. Unless you are reasonably handy with tools, philosophical about burning midnight oil, and able to do without the car for the days or weeks that it may be in the shop, you should not try this project. If you have the money, you can buy a factory-made diesel car or contract professionals to make the conversion for you. But if you yourself have done the work and the result is livable, you will have done something beyond the scope of most amateur and many professional mechanics, and you may be justly proud of your achievement.

THE LIMITS

Any swap is possible—one dragster transplanted a Merlin V-12 aircraft engine into a Mack garbage truck. But this kind of radical engineering is not for the home craftsman. He should approach the swap with the intention of doing as little violence to the vehicle as possible. Unless the swapper knows exactly what he is doing, major steering, suspension, or brake modifications are strictly verboten.

There are two major reasons for an engine swap going wrong: engine weight and engine size. Another 150 pounds will not make much difference to the average American sedan or pickup truck, but

Fig. 3-1. A Perkins NA-120 nestles under the hood of this '72 GMC pickup. With stock gearing, the truck reaches 80 mph and returns 15 – 20 mpg. (Courtesy RV World Magazine.)

double that weight and you will have an unguided missile that plows on turns, bottoms its front suspension, and locks its rear wheels during heavy braking. Although these problems can be alleviated, the vehicle will never behave as it did before the swap.

As well as being heavy, diesels are physically large for the power produced. An engine that is too big for the chassis can force you to reform the firewall, add a hood scoop, and do unsavory tinkering with the steering and suspension.

The moral: Deliberately underpower. Keep engine weight and size within the limits that the chassis can tolerate. At 55 mph you don't need horsepower: diesel torque will compensate for some of the loss.

Fig. 3-2. Plenty of underhood space make the Datsun the best bet among mini-trucks for diesel power.

THE VEHICLE

The candidate for diesel power should be a simple machine that has ample engine-room space and a suspension that can tolerate the additional engine weight.

Full-sized pickups are ideal (Fig. 3-1). Frame members are well defined, simplifying engine mounting. Suspensions are stronger than they have to be and underhood space is more than adequate. Because you expect less comfort from a pickup, you can change to a large, noisy engine to gain more power than the original engine had. Mentzer Detroit Diesel of Sparks, Nevada, is known throughout the West for its Detroit Diesel truck conversions.

Japanese mini-trucks are also good choices (Fig. 3-2). Engines are limited to four cylinders, displacing not much more than 150 cubic inches, but vehicle weights of less than 2500 lb allow adequate performance without turbocharging. Of the four makes that are imported—Datsun, Toyo Kogyo (Ford Courtier and Mazda), Isuzu (Chevrolet LUV), and Toyota—Datsun seems to be the most popular among swappers. These trucks are over-built, with rugged suspensions and adequate engine-bay room. Tony Capanna, owner of the Wilcap Co., in Torrance, Calif., is so impressed with the Datsun that he will soon offer a mail-order conversion kit. The engine to the kit is a four-cylinder Nissan, distinguished from the Chrysler version by a slightly larger displacement and a corresponding increase in torque and horsepower.

But then, the Toyota five-speed looks good....

The most difficult mini-trucks to repower are the Courtier and the Mazda. Their engine bays are some four inches narrower than other Japanese makes and their No. 1 cross-member can be a headache when installing engines with forward sumps. Rather than fabricate a new cross-member, most swappers are cutting out the old one, turning it upside down, and welding it several inches forward of its original position.

American passenger cars are another good bet. The most popular are the compacts, the class of vehicle represented by Chevy II, Nova, Vega, Valiant, Falcon, Comet, Maverick, Granada, and Gremlin (Fig. 3-3). These machines will accept a 50-60 hp oil engine transplant without suffering too much embarrassment on hills and freeway ramps. Engine bays are generously proportioned, since

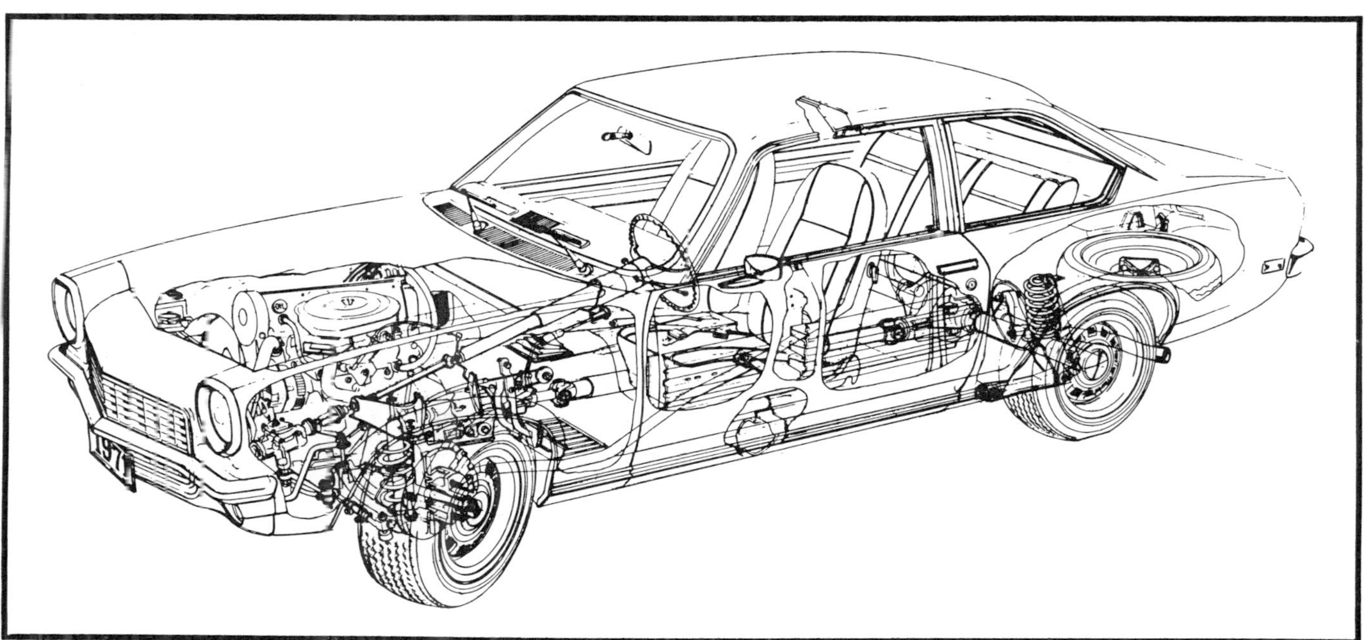

Fig. 3-3. The Vega makes good swap material: many engineless Vegas are around, vehicle weight is about right, and there is more room under the hood than one expects.

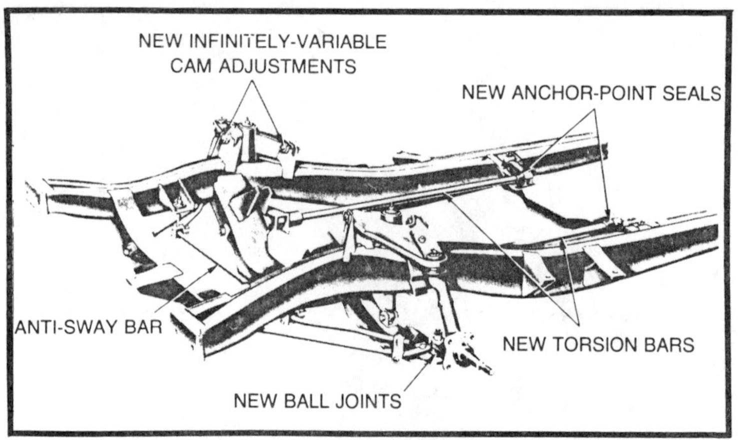

Fig. 3-4. Torsion bars help make the Valiant a desirable conversion.

their manufacturers dimensioned them for V-8 engines. Frames and suspension elements can easily support diesel weight, although adding more than 150 lb would justify a search for factory-air or preferably station-wagon front springs with heavy-duty shocks to match.

The Valiant seems to be the most popular subject for conversion. Chrysler-Nissan engines will bolt up to the manual or automatic boxes and front-end weight can be compensated for by additional preload on the torsion bars (Fig. 3-4).

Full-sized sedans, including a few luxury cars, have been successfully diesel-powered. Ray Todd, owner of Superformance Automobile Engineering, Ltd., of Oak View, Calif., recently installed a Ford 2402E diesel in his Continental Mark IV. This engine displaces 216 cubic inches and, with a Rajay 301B 60 turbochager, develops approximately 130 hp. At a steady 55 mph, Mr. Todd reports 28 mpg and very little engine noise. This conversion was built to convince the Ford Motor Company to install oil engines in their passenger cars. It was an exercise in high-level swapping and retained power-steering, air-conditioning, power-brakes, and the other amenities that Mark IV owners expect. A specially designed vacuum modulator feeds shift signals to the automatic gear box. Although the delegation of engineers from Dearborn was impressed with Mr. Todd's creation, this is not the conversion for an amateur to try. Nor is its 41 to 1 weight to power ratio worth his regretting my advice.

Small foreign sedans and sports cars are tempting, but there are enough difficulties to have kept most swappers away. Engine weight is the major problem. Original-equipment engines weigh half as much as the Mercedes-Benz 180D, which is the most readily available diesel suitable for these vehicles. Doubling the engine weight is not a good idea, particularly when vehicles like the Fiat already have marginal suspensions.

No attempt should be made to convert machines with unorthodox drive trains. Neither rear-engine rear-drive nor front-engine front-drive layouts lend themselves to diesel engines, particularly if the engine must lie transversely.

PLANNING

While it is possible to over-plan, whiling away months by anticipating eventualities which mostly never occur, reconnaissance and planning are important. It is, after all, easier to solve problems with pencil and paper than with a blow-torch and wrench.

The first consideration is space.

1. The distance between the top of the air filter and the bottom of the oil pan must be something less than the distance between the forward cross member and the hood.
2. The distance from the flywheel to the fan must be less than the distance from the bell housing to the radiator.

Within limits engine-room space can be stretched. A new, less intrusive cross member can be fabricated; the hood can be raised with a scoop; the radiator can be moved forward; or the fan can be discarded and replaced by an electric one mounted alongside the engine or in front of the radiator. But besides meaning more work, each of these expedients may create more problems.

The next concern is mating the engine with the drive train, chassis, and on-board systems (Fig. 3-5). I do not want to make the job sound harder than it is, but major interfaces are these:

1. engine block/bell housing
2. flywheel/clutch
3. crankshaft flange/pilot shaft
4. motor mounts/frame
5. water inlet and outlet/radiator and heater

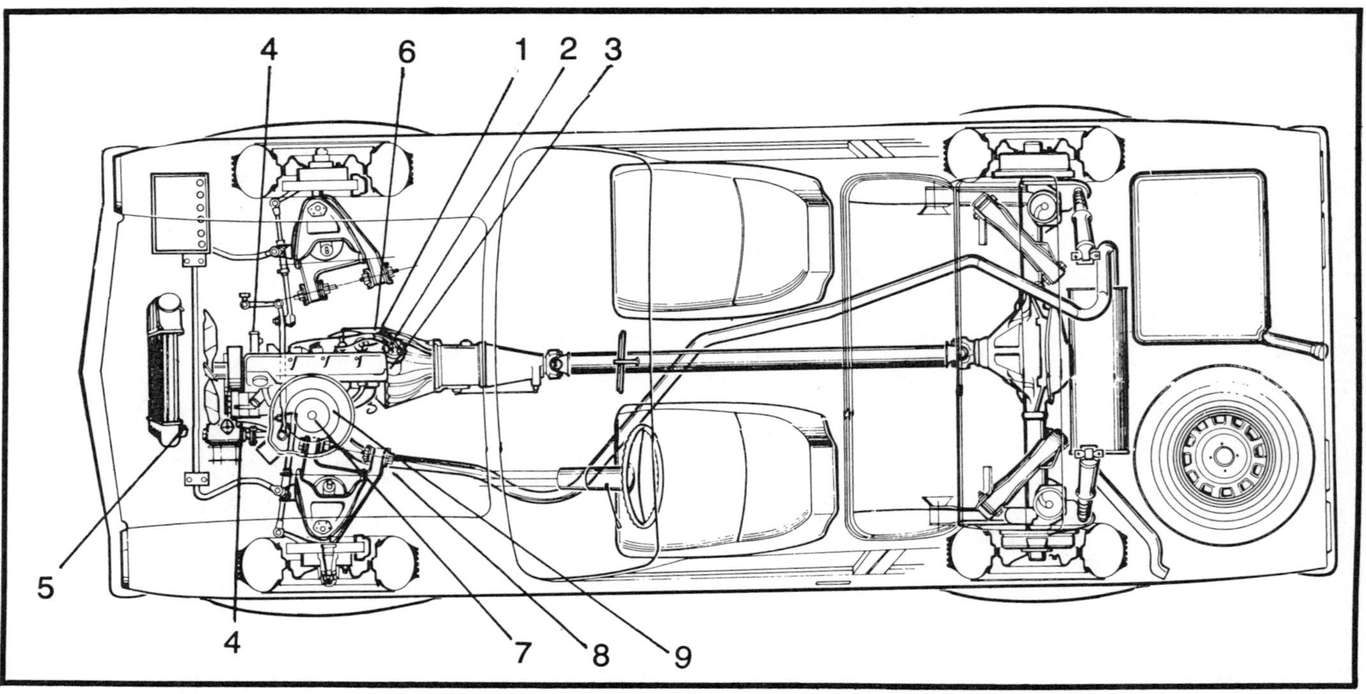

Fig. 3-5. Engine/chassis interfaces. Call numbers are referenced in the text.

6. starter motor/battery
7. exhaust header/exhaust pipe
8. pump rack/accelerator linkage

If you want a full complement of accessories, you must construct or adapt brackets and run drive belts for an air-conditioner pump, power-steering pump, and vacuum pump.

Some of these interfaces can be bridged easily. Even the most imposing one, the connection between the engine and the bell housing, can be a bolt-on proposition if you purchase an engine that fits your bell housing. Others, such as the coolant connections on the block and the fittings on the radiator tanks, may require outside help. It is a simple and inexpensive matter for a professional radiator-man to change the position of tank fittings.

It's a good idea to draw up a list of priorities, ranking tasks in the order in which they must be done. The first priority is to acquire or create a vehicle that can be mated to the engine. There is little that you can do to make the engine conform; the burden of modification falls on the car. Next, take the suitable engine and overhaul it as much as necessary to make it road-worthy. Concerning the details of the swap, I cannot give a dogmatic, hard-and-fast order of priority. So much depends upon your choice of vehicle and engine, your tools and your skills. But the main thing is to mate the engine with the driveline. All spacial relationships will be fixed by this connection. Make or adapt motor mounts to secure the front of the engine and then make the remaining hookups, tackling the most difficult jobs first.

TOOLS AND PARTS

You will need the usual mechanic's tools and a drill motor, and you should have access to arc- and gas-welding equipment. A fairly simple swap can be done without welding but the work will go more slowly and you will have less freedom in solving the fabrication problems.

Parts are readily available for most diesel engines sold in this country. You may find that parts for the same engine are stocked by industrial, marine, refrigeration, and truck outlets. Alternatively, you may deal directly with the factory or importer.

Most of the heavy metal such as mounts for engine and transmission, bell-housing adaptors, and accessory brackets must be fabricated for the swap at hand. Although there are "universal" motor and trany mounts available from West Coast specialty houses, and although commercial swappers do sporadically offer kits, you had better expect to be self-reliant. The field is still new. No one engine/vehicle combination has yet proved to be overwhelmingly popular. Until there is a standard swap, no large manufacturer will be interested in mass-producing the necessary hardware.

Although a clean swap requires only minor modifications to the vehicle, it is nice to know that Detroit has engineered heavy-duty brake and suspension kits for most sedans. Front coil springs on cars with factory air are heavier than stock, and can compensate for an additional hundred pounds of engine; station wagon and police special springs can support even heavier loads. High-performance brake-pads and brake-shoes are available from the dealer or brake manufacturer.

Engine accessories and instrumentation are the finishing touches to a swap, giving the result the elegance that we expect from diesel power. This gear is expensive when new, but a little poking around through truck wrecking yards may uncover such goodies as pyrometers for monitoring exhaust temperature, air flaps for emergency engine shutdown, cold-start kits, audible oil-pressure and coolant-temperature alarms, or thermostatic radiator shutters.

While we're on the subject of junkyards, I do not think that the time you spend in one is ever wasted. A junkyard is a museum of automotive technology. By studying the hulks, you can understand how factory engineers solved the same problems that you are facing. And you can often use the very same hardware.

In conclusion, choose a basically simple vehicle with a generously sized engine bay, plan carefully, and research your local wrecking yards.

Buying An Engine

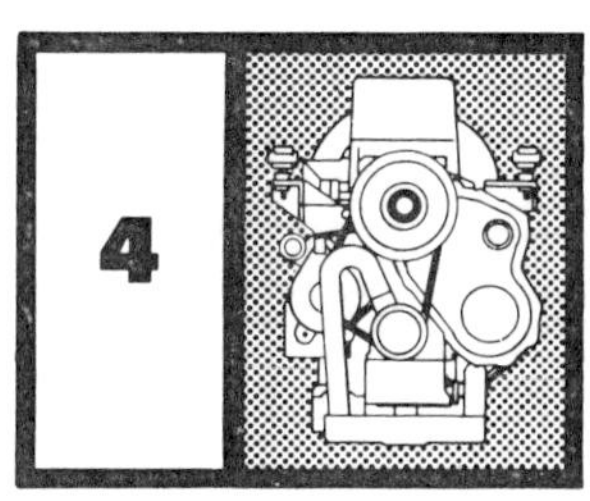

Before you buy an engine, measure the engine compartment in all three dimensions (Fig. 4-1). Determine:

- Horizontal distance from the firewall to 1 inch aft of the radiator
- Vertical distance from the top of No. 1 cross member to the underside of the hood
- Transverse distance between the inner fenders, or splash panels.

These preliminary measurements can be made with a yardstick for most swaps. The larger the engine is in proportion to the available space, the more carefully you will have to measure. If things appear tight, a helper and a flexible steel rule you can snake around obstacles are useful. It is necessary to pull the engine when absolute accuracy is required.

Make special note of the position of the front cross member, since it is the logical attachment point for the motor mounts and will determine the location of the oil sump (Fig. 4-2). Check for possible interference at the steering links, shaft, and box, as well as the upper suspension elements and master cylinder (Fig. 4-3).

NEW ENGINES

New engines can be purchased with vacuum pump, automotive filters, and, in most cases, with compatible bell-housing bolt-circles.

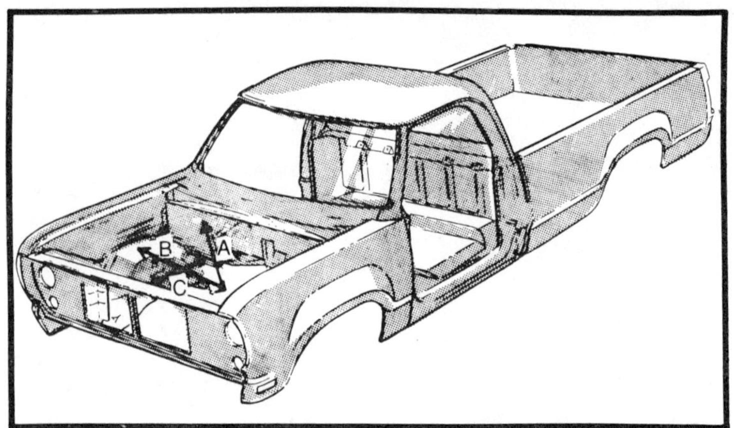

Fig. 4-1. The height of the engine compartment is the distance from the top of No. 1 cross member to the underside of the hood (A); the critical width is the distance between the inner fenders (B); the effective length is the distance from the bell housing to 1 inch aft of the radiator (C).

CHRYSLER-NISSAN

Chrysler-Nissan engines are adapted for automotive, marine, and industrial roles. They are made by Nissan at Marysville, Michigan, and marketed by Chrysler. Two basic engine types are offered: four- and six-cylinder.

The CN4-33 is an inline-four, displacing 132 cubic inches (Fig. 4-4). Rated at 48 hp, it develops a maximum of 61 hp at 4000 rpm (Fig. 4-5). The engine features swirly-type precombustion cham-

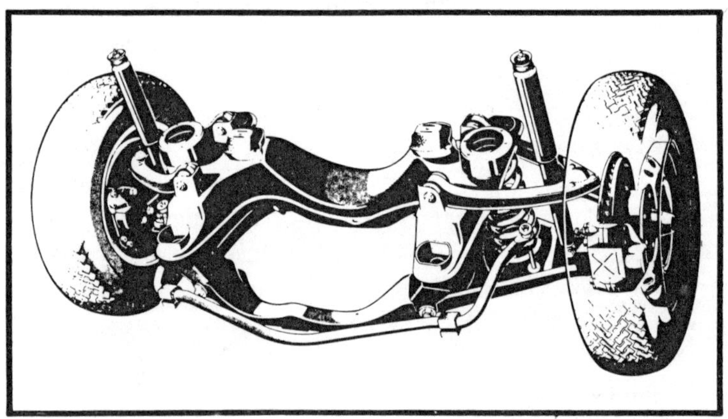

Fig. 4-2. Mercedes-Benz sumps cradle between the crossmembers, making things difficult for engine swappers.

bers, replaceable dry cylinder liners, five-ring pistons, and a Diesel KIKI inline pump which looks as though it were built on the old Bosch license. In spite of its weighing approximately 474 lb with flywheel and accessories, the 4-33 is a popular swap for compact cars.

The CN6-33 develops 92 hp from 198 cubic inches (Figs. 4-6 and 4-7). All-up weight is 662 lb. This engine is the choice of International Harvester and most swappers. It is virtually bullet-proof—Chrysler engineers have not seen a valid warranty claim for lower-end failure in five years; relatively quiet; and reasonably economical.

CHRYSLER-MITSUBISHI

Chrysler also markets two Mitsubishi engines. The CI 641-100, which is the 6DR5 of Mitsubishi terminology, is a six-cylinder, four-cycle engine developing 110 hp at 3700 rpm (Figs. 4-8 and 4-9). The engine displaces 244 cubic inches and weighs in at 790 lb with accessories, flywheel, and flywheel-housing. The SAE No. 3 bell-housing mates to many American transmissions. The Mitsubishi, like the Nissan, features replaceable dry sleeves, five-ring pistons,

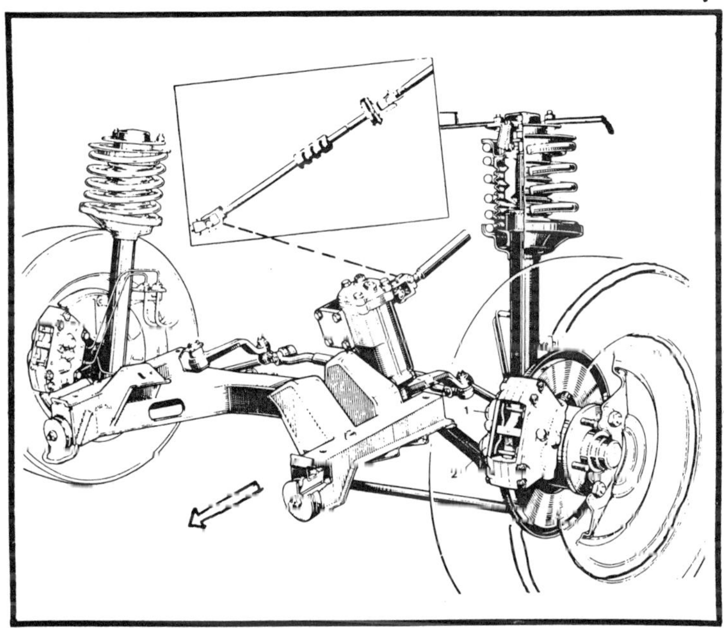

Fig. 4-3 Steering components may intrude into the available space.

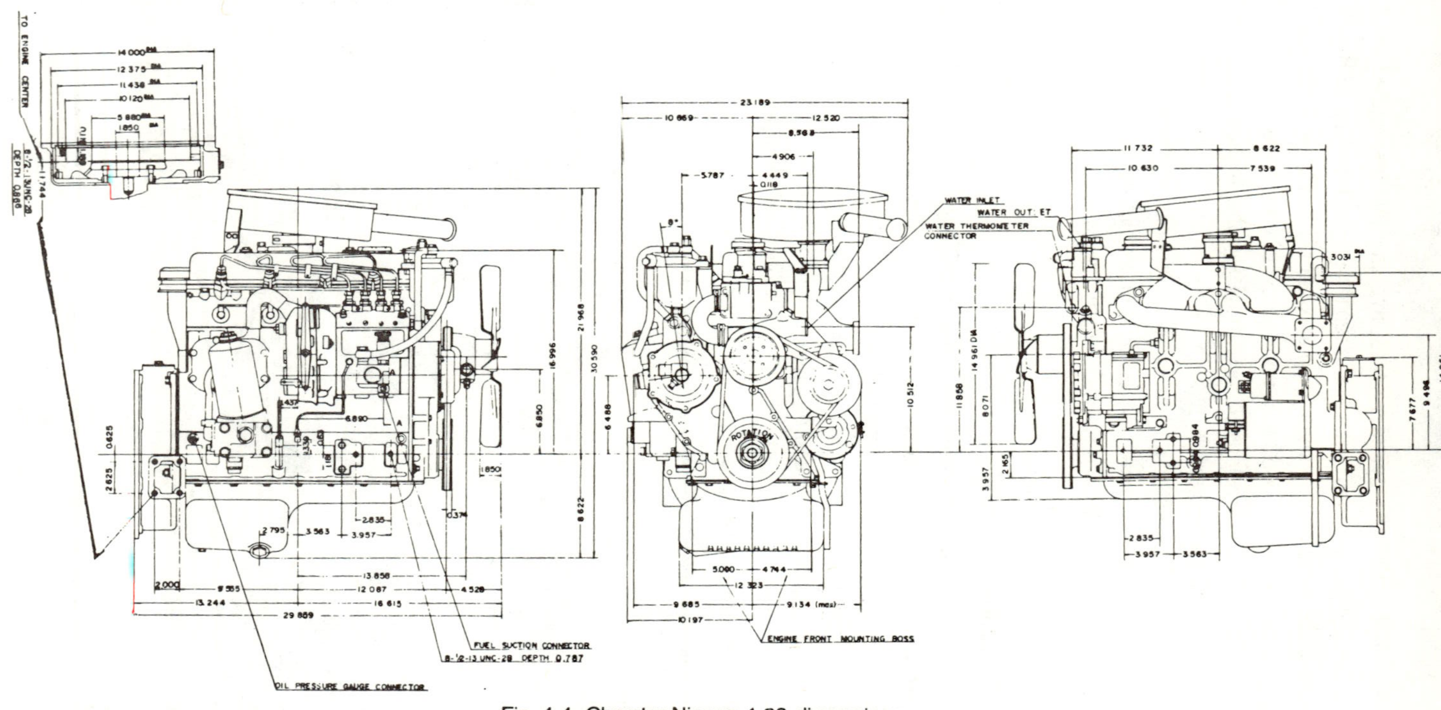

Fig. 4-4. Chrysler-Nissan 4-33 dimensions.

centrifugally governed Bosch-type injection, and a drop-forged steel crankshaft riding on seven main bearings.

Rumor has it that the Mitsubishi will become *the* Chrysler diesel when the current contract with Nissan expires, but Chrysler spokesmen point out that there is room for both in the market and in their line: the CI 641-100 displaces 46 cubic inches more than the Nissan and outweighs it by 128 lb. On the other hand, Chrysler has had quite a history of littering the land with such orphaned imports as the Simca and the Cricket.

The Chrysler-Mitsubishi CI 655-100 is a 331 cubic inch truck engine, developing 130 hp from its six cylinders (Figs. 4-10 and 4-11). Governed speed is 3150 rpm. At 940 lb the CI 655 is no lightweight, but its awesome torque characteristics make it suitable for heavy pickups and large sedans.

DETROIT DIESEL

Detroit Diesel's three-cylinder block is known as the 3-53 (Fig. 4-12). It displaces 159 cubic inches and develops 91 horsepower in

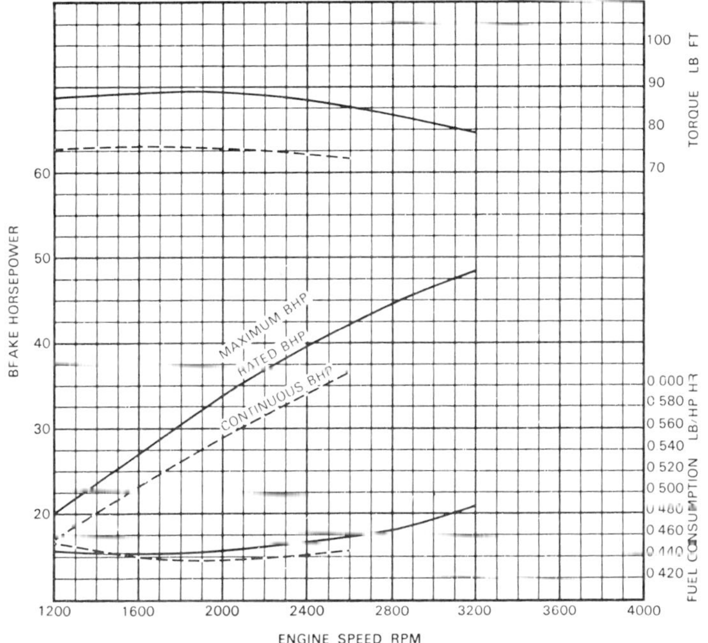

Fig. 4-5. 4-33 performance curves.

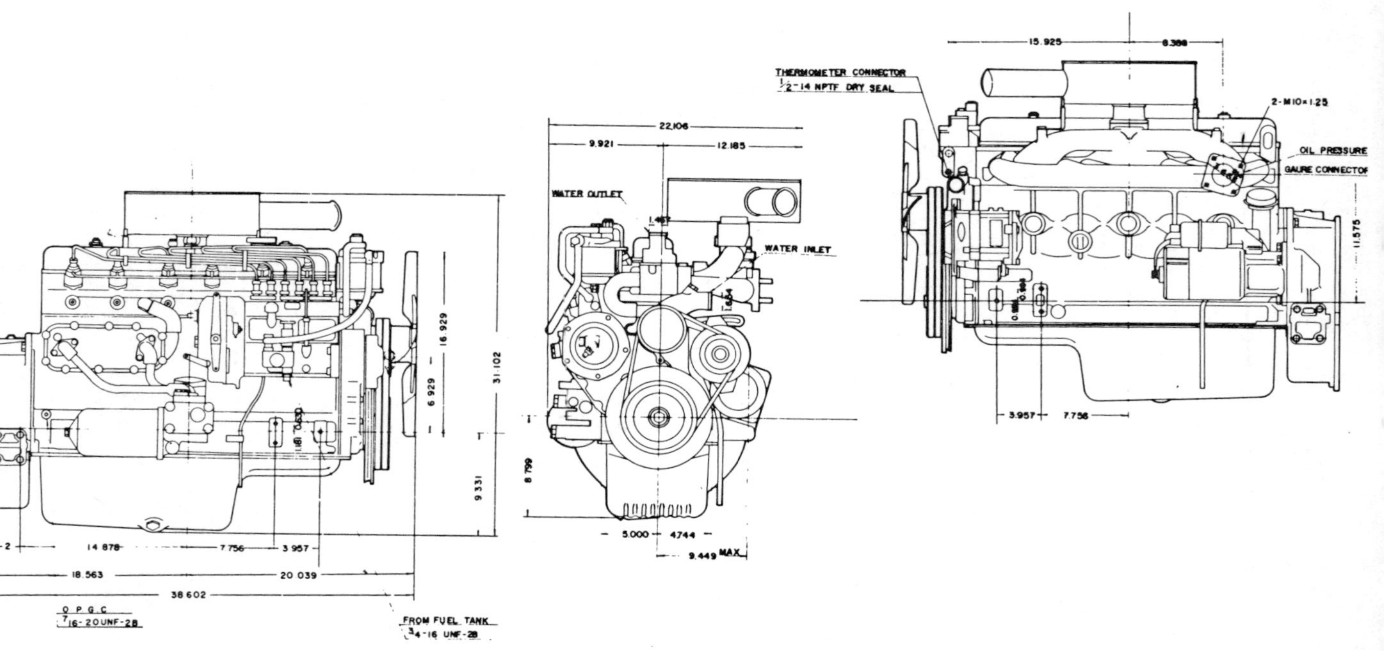

Fig. 4-6. Chrysler-Nissan CN6-33 dimensions.

the 4-valve configuration. It measures approximately 33 inches long, 35 inches high, and 27 inches wide. Dry weight, including flywheel, is 965 lb. The flywheel housing and flywheel are built to SAE No. 4 specifications.

The 4-53 is the same engine with another cylinder. It displaces 212 cubic inches and develops 123 hp with the optional four-valve head and N45 injectors (Fig. 4-13). Roughly 39 inches long, 35 inches high, and 26 inches wide, the engine weighs 1110 lb, dry.

Detroit Diesels are two-cycle engines, running with wet crankcases and scavenged by positive-displacement blowers. Much loved by truckers and immortalized in honky-tonk ballads, the Detroit moan is heard all over this country. A Detroit-powered pickup truck is the ultimate macho trip. But listen to a Gimmy run before you do anything so rash as to install one in a passenger car.

DEUTZ

Made in Cologne, Germany, the Deutz engines are taking an increasing share of the American market (Fig. 4-14). Like the

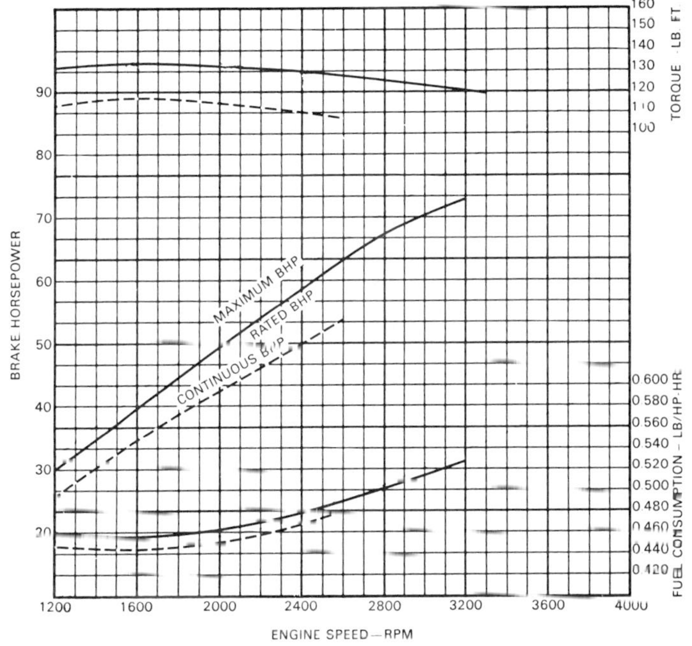

Fig. 4-7 CN6-33 performance curves.

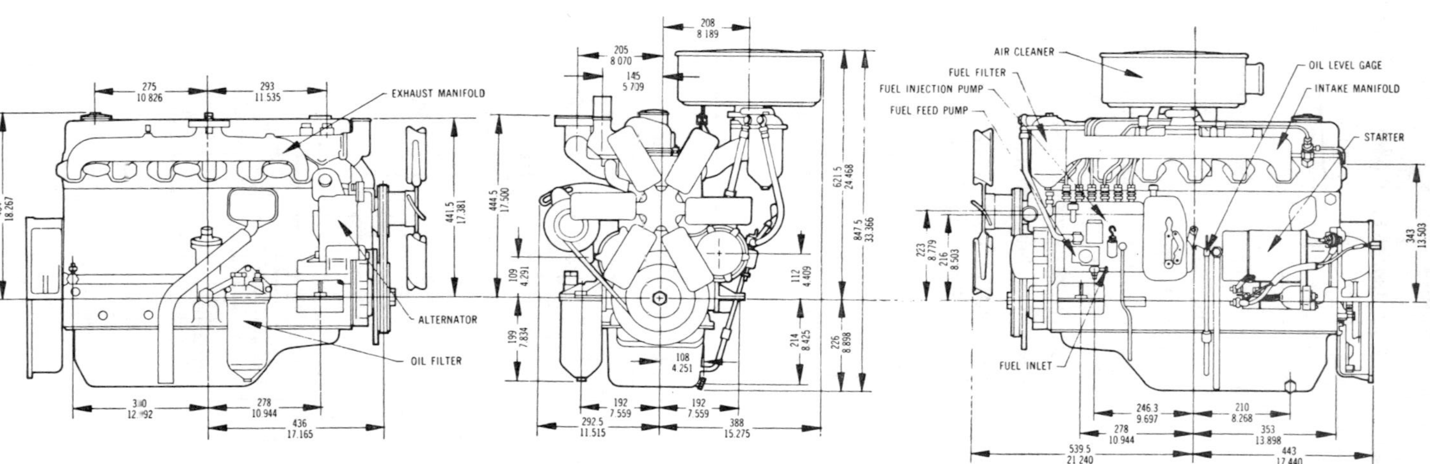

Fig. 4-8. Chrysler-Mitsubishi CI 641-100 dimensions.

Czechoslovakian Tatra, these engines are air-cooled and built on a rigid modular pattern. Pistons, rods, cylinder barrels, heads, and valve gear are interchangeable within a given series, whether the engines have one, five, or twelve cylinders. Their air cooling has a number of advantages:

- Engine weight is reduced. The conventional water jacket weighs almost as much as cooling fins, and here there is no water pump and radiator.

- Reliability improves. The cooling system accounts for 20% of liquid-cooled engine failures.

- Durability improves. Air-cooled engines warm up faster and operate at slightly higher temperatures. Tests have shown that air-cooled cylinder bores last longer than those that are water-cooled.

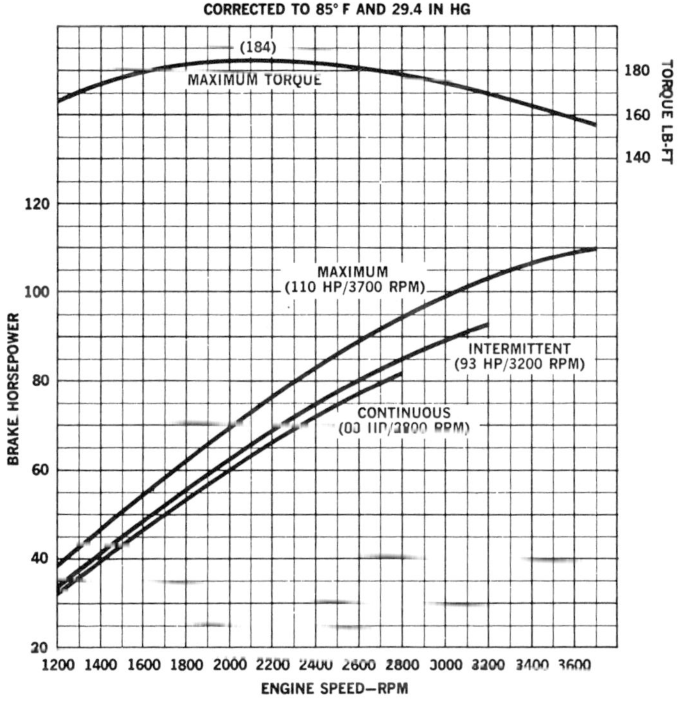

Fig. 4-9. CI 641-100 performance curves.

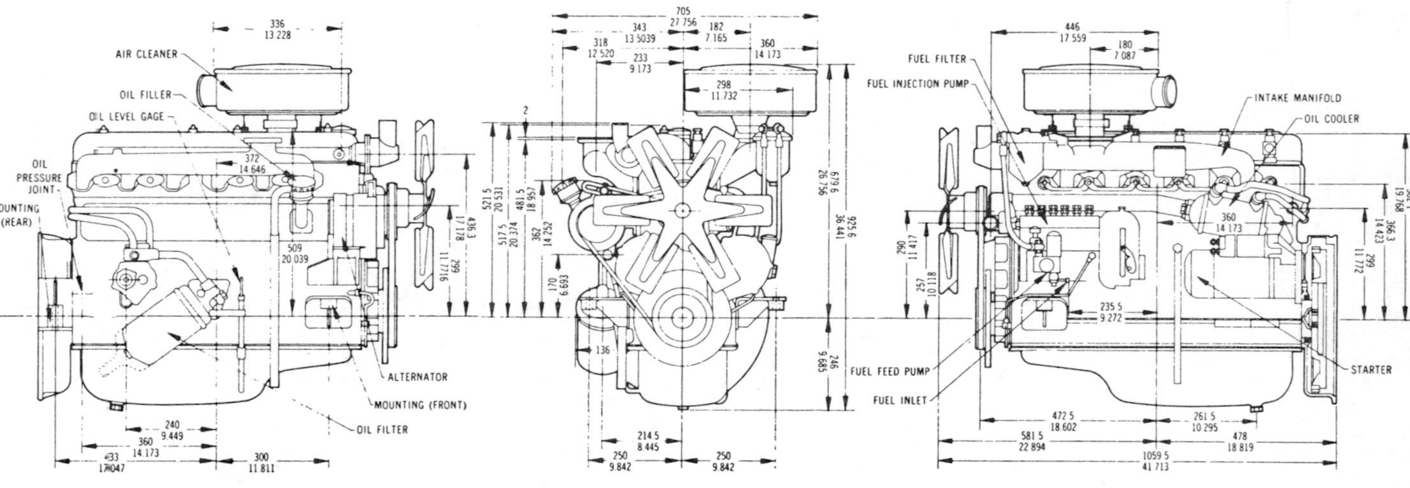

Fig. 4-10. Chrysler-Mitsubishi CI 655-100 dimensions.

- Maintenance is simplified by the cylinder arrangement. Since each cylinder is finned around its entire circumference, cylinder barrels and heads are detachable. A mechanic can change the piston, rings, and valves on an individual cylinder without dismantling the entire engine.
- Emissions are lower because of the higher cylinder temperatures.
- Installation is simplified.

Nor are air-cooled engines necessarily noisier than liquid-cooled types. Cooling fins do not radiate sound. The noise output of a single-cylinder test engine was measured at the Tatra Works. The fins were milled and a cooling jacket was welded into their place. The engine made just as much noise as before. An experiment at NSU corroborated the results: plastering the fins over with concrete had

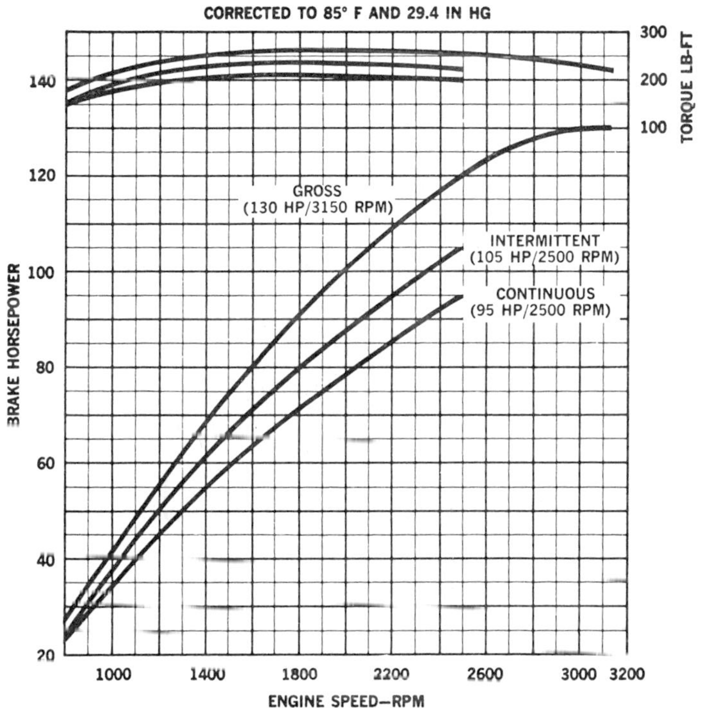

Fig. 4-11. CI 655-100 performance curves.

Fig. 4-12. Although the 3-53 has only three cylinders, the torque pulses match those of a four-cycle six.

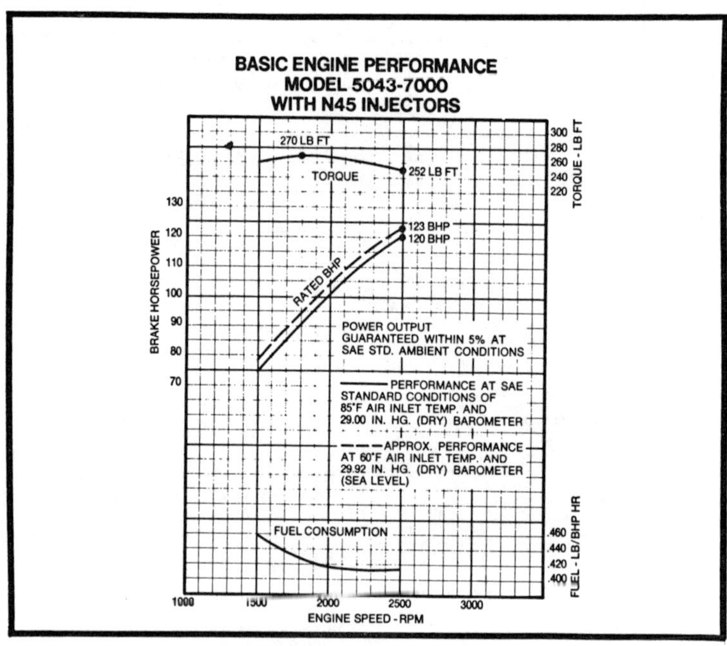

Fig. 4-13. 4-53 performance curves.

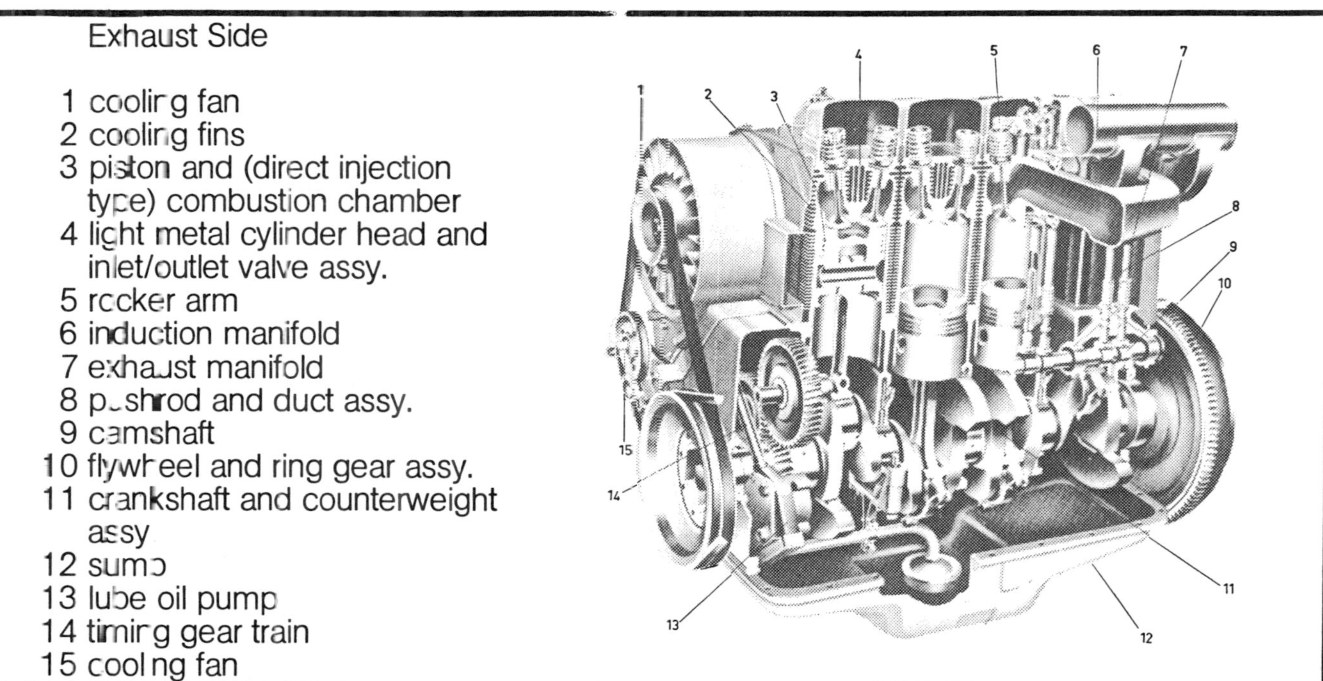

Fig. 4-14. The Deutz F4L 912 in cutaway view. Indirect injection is optional.

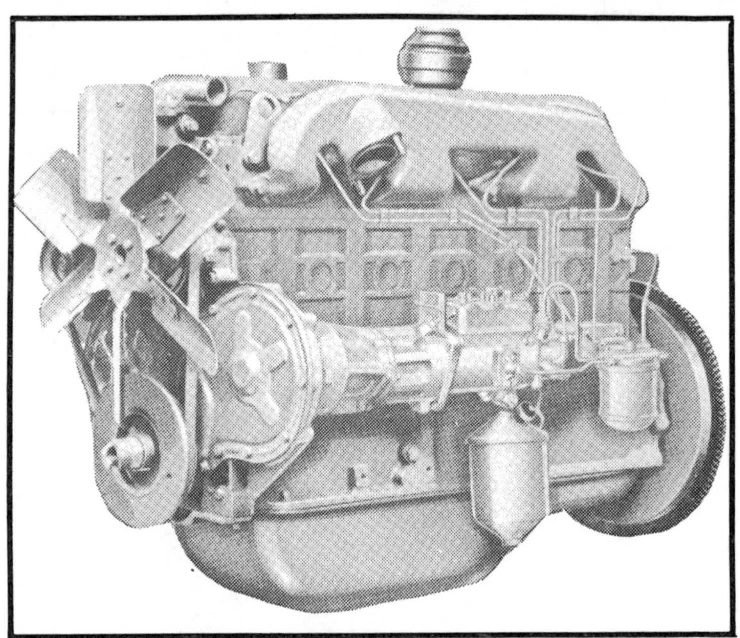

Fig. 4-15. The GM-Bedford Model 330.

no effect upon the noise level. L.O. McClure, a Deutz dealer in Bakersfield, California, installed one of these plants in a pickup truck with satisfactory results, but he cautions that the job is not a weekend project.

GM BEDFORD DIESEL

GM Bedford engines are widely known in Europe where they are used in construction and trucking. The model 330 is a six-cylinder, four-cycle design, displacing 330 cubic inches (Fig. 4-15). It is rated at 100 hp at 2600 rpm and weighs 1015 lb, including flywheel and accessories. The block is drilled for an SAE No. 2 bell housing. A CAV distributor-type pump provides the fuel. The 330 is approximately 40 inches long, 24 inches wide, and 34 inches high. The oil is sumped forward in the pan, a situation that could cause cross-member interference.

ISUZU

Isuzu Motors, Ltd., is the oldest automotive manufacturer in Japan. Established in 1916 to license-build Wolseley cars and trucks, by 1936 the firm had developed a line of air-cooled diesel engines, trucks, buses, and off-road vehicles. After the war Isuzu continued to build trucks and engines. In 1966 the first new passenger cars again came off the Fujisawa assembly line. Under contract with General Motors, Isuzu builds the LUV pickup and the Opel. This preamble is necessary because few Americans have heard the name of the manufacturer of two of the best light diesels available.

The 4-BB1 (Fig. 4-16) and the 6-BB1 (Fig. 4-17) are, respectively, four- and six-cylinder versions of the same engine. As you can see from Fig. 4-16, the 4-BB1 will fit in a space 19 × 25 × 33 inches. The engine weighs 732 lb and develops 90 hp at a conservative 3300 rpm. It will bolt up to SAE No. 3 bell housing. The 6-BB1 displaces 328.5 cubic inches, develops 128 hp, and weighs 992 lb.

Both employ inline, centrifugally-advanced fuel pumps. Both can be fitted with direct-drive power steering pumps as an option. The 4-BB1 has the pontoon sump shown in the drawings as standard equipment; a rear sump is available upon request.

PERKINS

An old-line English manufacturer, Perkins has built engines for Land Rovers, Checker cabs, Dodge and White trucks, IHC school buses, and Bombardier all-terrain vehicles. While some manufacturers consolidate production into a handful of modular engines, adding or subtracting cylinders as output requirements vary, Perkins prefers to tailor their engines to the use. The range of Perkins' line is staggering, including three-cylinder lift-truck engines, and horizontally-opposed and V-8 marine plants.

Their NA-70 is an inline-four, displacing 153.9 cubic inches (Fig. 4-18), churning out 70 hp at 3600 rpm, and weighing only 590 lb with accessories. The largest automotive engine in the series displaces 605 cubic inches and delivers 210 hp. As typical for Perkins' products in general, these engines employ dry cylinder liners, molybdenum alloy connecting rods, and Roosa-Master pattern pumps.

USED ENGINES

Nearly every person who contemplates a diesel conversion will at some time or other entertain the idea of utilizing a used engine.

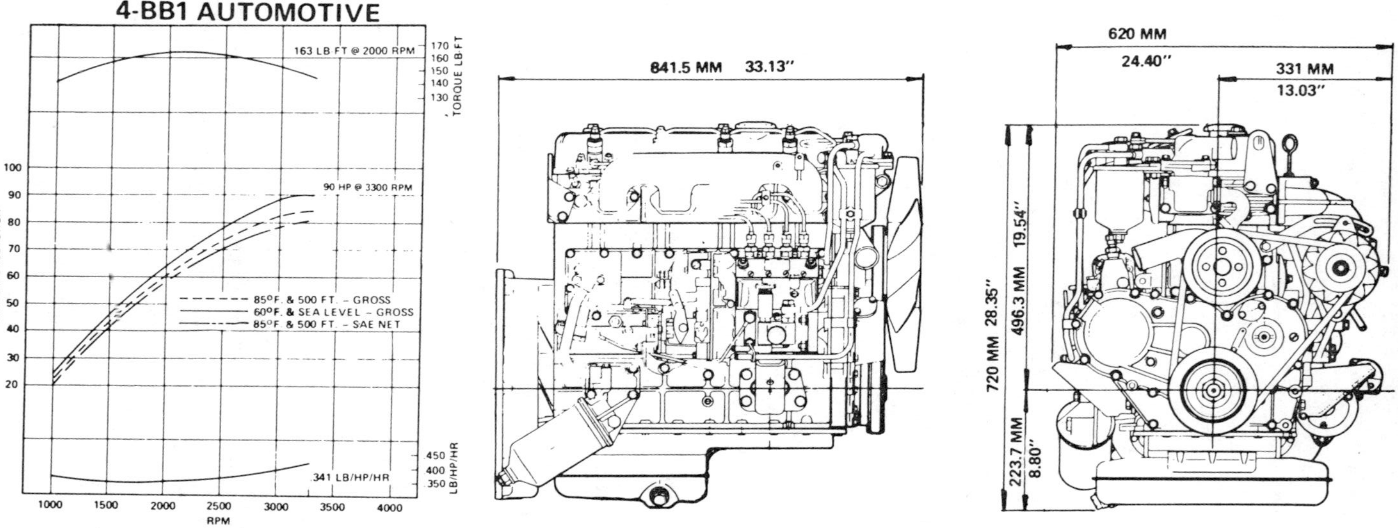

Fig. 4-16. The Isuzu 4-BB1 is a compact engine designed for automotive applications.

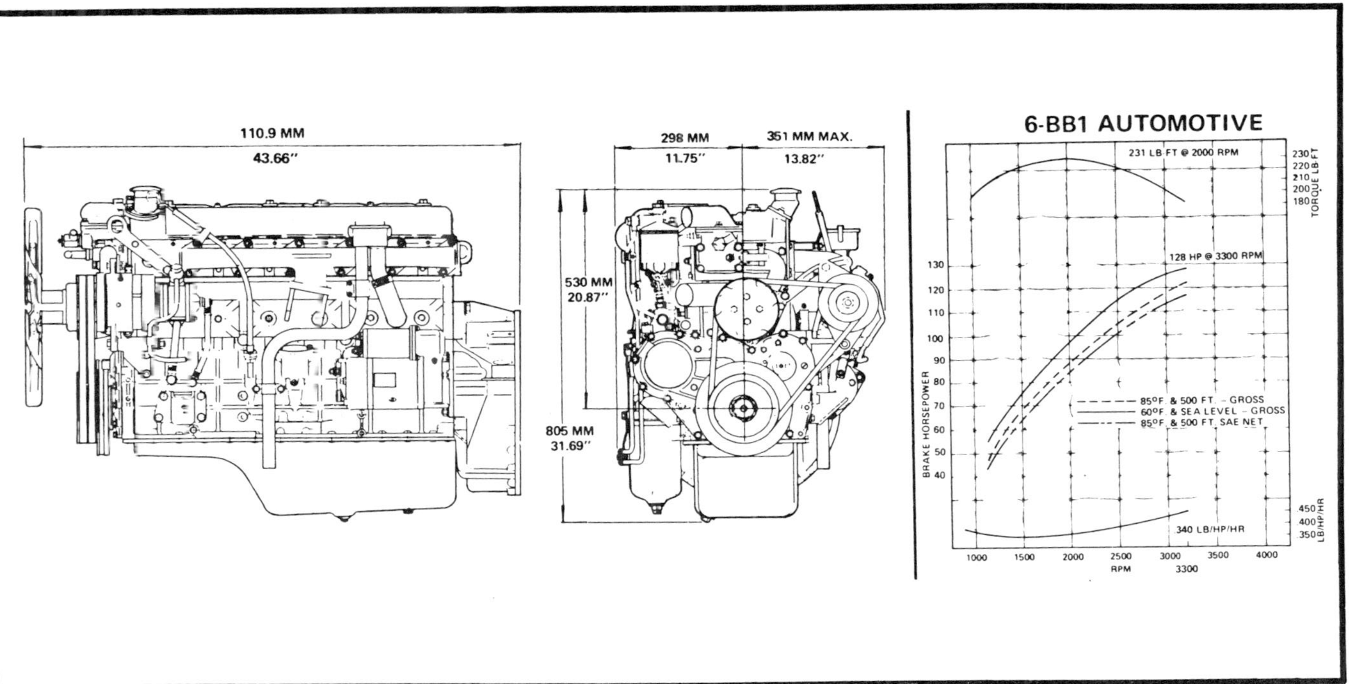

Fig. 4-17. The 6-BB1 is intended as a truck engine.

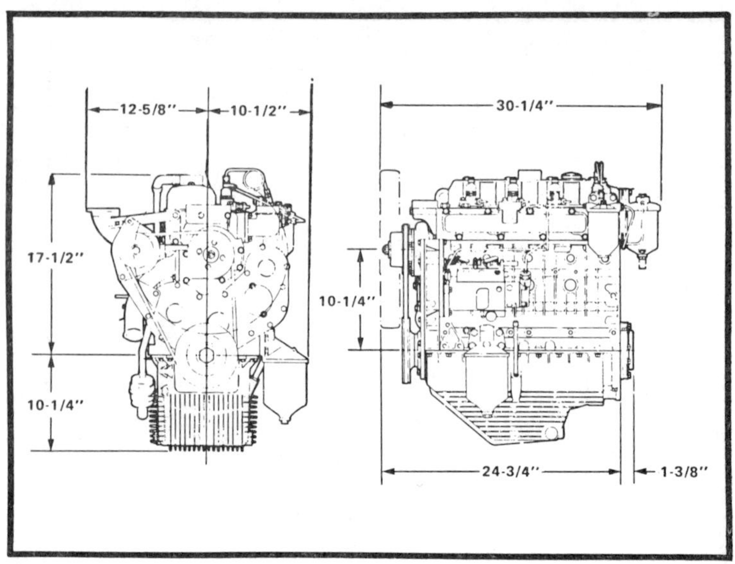

Fig. 4-18. Perkins NA-70 dimensions.

What's Available

Used engines are in demand by farmers, factory owners, truckers, small-boat skippers, and impecunious swappers. The high road is buying a factory remanufactured engine, the middle road buying a rebuilt engine from a local shop, the low road is buying an engine from a junkyard "as is."

There is little to say about the first two options. You pay your money, get your warranty, and hope for the best. It helps if the people who reworked the engine are reputable. It helps even more if the engine was in fairly good shape to begin with.

Used engine sources are automobile and truck wrecking yards, boatyards, dealers, and repair shops. Prices will vary according to the local interest in diesel power. In the Cumberland Valley, complete as-is engines go for $150-$200.

The swap will be simplified if you buy a complete engine/transmission assembly, as well as miscellaneous hardware, from a diesel car. In practice, the choice falls upon Mercedes-Benz because other diesel sedans are almost non-existent in junkyards. The 220-D and later series are the more desirable Mercedes. These engines are substantially more powerful than their predecessors and take more kindly to turbocharging.

Other good choices are the Mercedes OM 636 and OM 621. These plants are industrial versions of the 180 and 190 (Figs. 4-19 and 4-20), respectively. The OM 636 is really available in truck junkyards, since tens of thousands of them power Thermo-King reefer plants. Governed at 2100 rpm, a properly maintained OM 636 should outlast the semitrailer it cools. I have disassembled one of these engines with 4800 hours on the clock. The rod and main-bearing clearances were within factory wear limits, and machining marks could still be seen on the piston flanks.

This near-immunity to wear extends only to speed-sensitive parts. You can expect valve, injector, and possibly fuel-pump problems on high-hour engines.

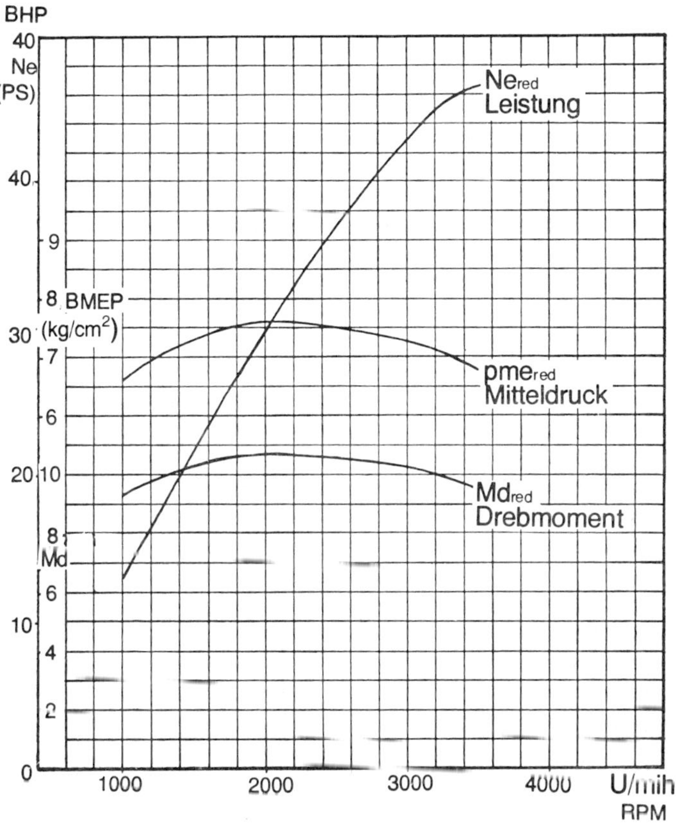

Fig. 4-19. Mercedes-Benz 180D/OM 636 performance curves. DIN horsepower ratings are conservative by American standards.

Fig. 4-20. The 190D was the first Mercedes-Benz diesel to use an overhead camshaft.

Nor are industrial engines, even those that are cousins to automotive plants, a cinch to adapt. The exhaust header may point skyward; cooling and oiling systems may have multiple outlets to service the driven machinery; flywheels are almost always incompatible with automotive clutches; or as in the case of the OM 636, the oil sump must be replaced by something less gargantuan.

Tractor and construction-equipment engines can be excellent choices, particularly if they had been designed with an eye to automotive use. British-built Ford engines are good examples of automotive plants disguised as tractor engines. On the other hand, Case engines with their paired cylinder castings and massive crankcases belong in a tractor and nowhere else. The same is true of the early, glow-plugless International Harvester engines.

Marine and truck engines are good raw material for a swap, if you can find one that is not decrepit. Truck engines can easily total half a million miles before they are junked. Rebuilding an engine that has seen so much service may be futile and is always expensive whatever the outcome. Marine engines do not suffer the speed and load fluctuations of truck plants, but are operated full-bore for hours or days on end. You can expect to see severe cylinder, piston, and crankpin wear on even the cleanest examples. Of course, some modification of the cooling system will be necessary to adapt these engines for road use (Fig. 4-21).

As a group, the worst buys are engines that have powered industrial lift-trucks. Don't be talked into one. No service is more demanding. The industry estimates that each hour on the clock is the equivalent of 35 miles' distance for a road-going vehicle. The engine is subject to constant acceleration and deceleration, and loads are severe and abrupt. As the truck ages, maintenance becomes increasingly difficult. It is a battle to keep the old truck operating throughout the entire shift.

Determining Engine Condition

Buying an engine without first listening to it run is foolhardy, but sometimes necessary. You can't be overly choosy about a $100 diesel, considering that a new fuel pump would cost seven or eight times as much, but you should at least bar the crankshaft through a full revolution to determine that the engine is free. No movement might mean a broken rod, or it might simply mean that the piston rings have rusted solidly into the bore. Movement through part of a revolution is more serious, almost always heralding a first-class calamity. If possible, pull the cylinder head. Check the condition of the bores, valves, and piston tops as described in Chapter 6. And, while the head is still off, bring each piston successively up to top dead center and rock the crankshaft a few degrees. The pistons should respond obediently. Any noticeable lag means bearing and/or crankpin wear.

Another test deserves special mention. It is possible to determine engine condition with scientific exactitude by spectroscopic analysis of lube oil. Since the oil circulates the wastes generated by the engine, it will contain metal fragments from any disintegrating parts.

Oil testing was pioneered by the railroads during the period of transition between steam and diesel. Railroad mechanics were still unfamiliar with compression-ignition engines, and some running-check on engine condition was needed. These early tests could detect bearing-failure, but the Navy, principally at its testing station

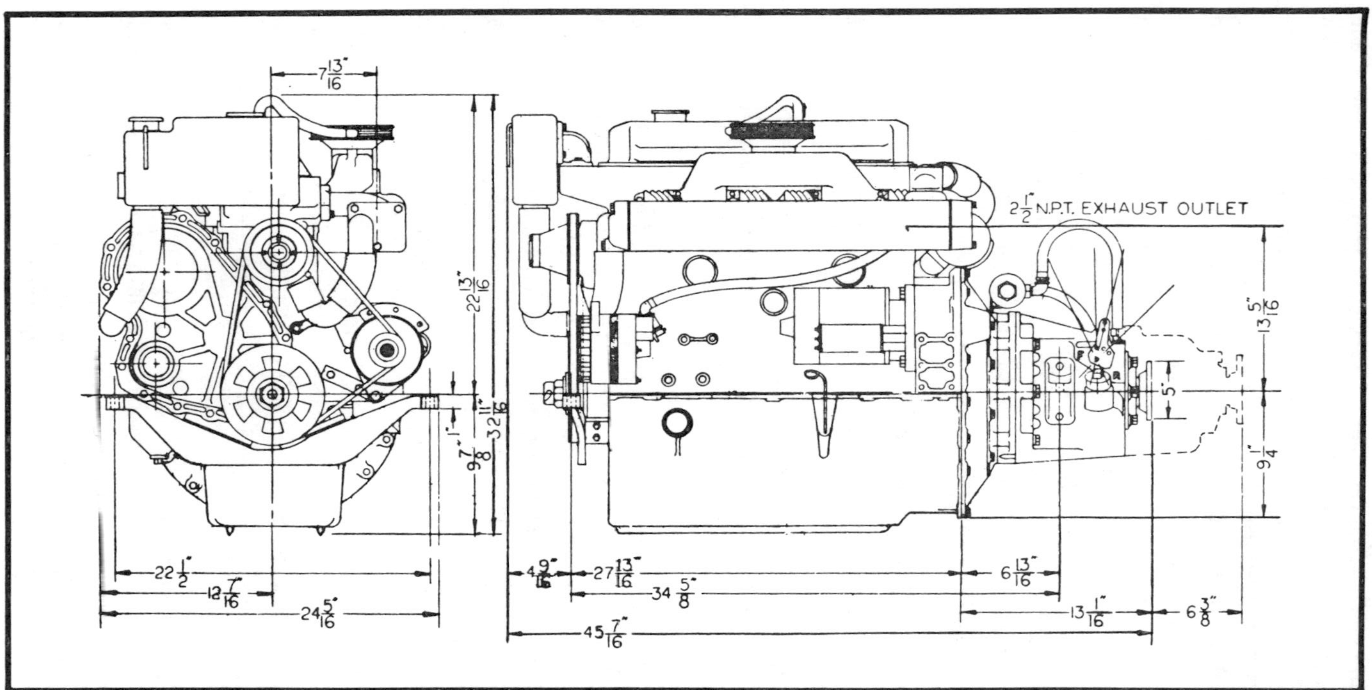

Fig. 4-21. The Barr 254 is a marine version of the Ford four-cylinder. Header tank and marine transmission are superfluous for our purposes.

in Pensacola, Florida, made oil testing a forecasting instrument. By the mid-50s Navy engineers could predict engine failure with chilling certainty. More than one plane was grounded, often with the engines warming for takeoff, by an emergency phone call from the testing station. Since then spectroscopic analysis has been increasingly used by private industry.

The significant trace-materials found in lube oil and the sources of these materials are in Table 4-1. In addition, silicon and aluminum traces indicate a failure of the air filter or severe leaks in the intake tract. Other metallic elements such as boron, zinc, and calcium are present as oil additives, but when found in high concentrations should be considered contaminants.

Spectroscopic analysis involves vaporizing the oil sample in an electric arc. Part of the characteristic signature of each chemical element is that it gives off light at a single frequency. The flash from the vaporized sample is directed to a prism that sorts the light by frequency. Photocells record the intensity of each component, allowing the contaminant level to be read directly from meters or from computerized printouts. Modern spectrometers can detect traces as fine as one part per million.

A typical oil test identifies 16 substances and may be repeated several times. This raw data is turned over to an analyst who, on the basis of other information about the engine such as its history and service profile, will make the final report. If he discovers that the engine is about to blow, he will call you by phone. This may sound frightfully expensive, but Analysts, Inc., (820 E. Elizabeth Ave., Linder, N.J. 07036) will do the job for less than $15.

Table 4-1
Lube Oil Analysis

Trace Material	Source
Lead	bearings
Silver	bearings
Tin	bearings
Aluminum	bearings, rods
Nickel	bearings, valve
Copper	bearings, bushings
Iron	cylinder bores, piston rings
Chromium	rings

Table 4-2

The Idle Test

Color of Smoke	Indicates	Cause
Blue	Excessive oil consumption	Usually associated with worn rings and cylinder bores, but it may also involve leaking valve seals or pullover from an overfilled sump.
White	Incomplete combustion	Water in fuel (which may not be visible). Late timing. Excessive or unequal fuel injection. Insufficient compression.
Black	Incomplete combustion	Some smoke normal during acceleration. Persistent smoke, however, can mean early injection, faulty injector spray patterns, insufficient compression pressure, or restrictions in the air intake.

Crankcase oil can be used for this test, but for best accuracy, tap into the circulating oil supply at the filter. Two or three ounces will be a large enough sample.

Assuming that the engine can be cranked, a compression test can give some notion of the condition of the cylinders, rings, and valves. A diesel compression gauge mounts in place of one of the injectors. The compression ratio and test conditions both affect the gauge readings. The compression ratio may vary from 17 to 23 in automotive-type engines; and you must consider such variables as battery charge, ambient air temperature, oil viscosity, or pressure drop across the air filter. You can expect between 15 and 30 kg/cm^2 (215 – 430 psi) from a reasonably healthy engine. No cylinder should vary by more than 20% from the average of the others.

Low readings can mean a holed piston, a blown head gasket, leaking valves, or worn rings. If it is the rings that are at fault, a few cc's of oil poured into the cylinder through the injector port will cause a dramatic rise in compression.

If the engine can be started, allow it to idle until it reaches operating temperature. Pull out the rack and observe the color of the exhaust. Table 4-2 relates smoke color to its probable cause.

The Fuel System 5

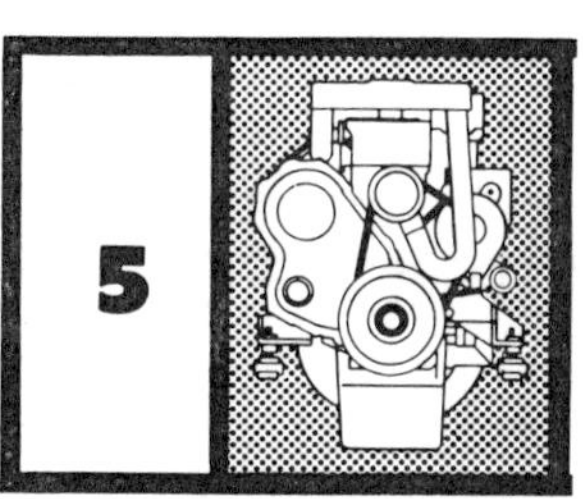

The fuel system is the most sensitive and by far the most complex of all diesel systems. Engine malfunctions usually originate here.

Fuel systems have these basic parts:

- Tank
- Lift, or low pressure, pump
- Filters
- Priming pump
- Injection pump (separate or combined with the injectors)
- High pressure lines
- Injectors
- Governors

THE PUMP

The three basic fuel systems used in small engines are most readily distinguished from each other by their different pumps. These are the inline pump, the distributor or "rotary" pump, and the injector pump.

The inline pump was pioneered by Robert Bosch in Germany and it survives on Chrysler-Nissan, Ford, and Mercedes-Benz engines. Figure 5-1 is a typical example. Fuel is drawn to the injection

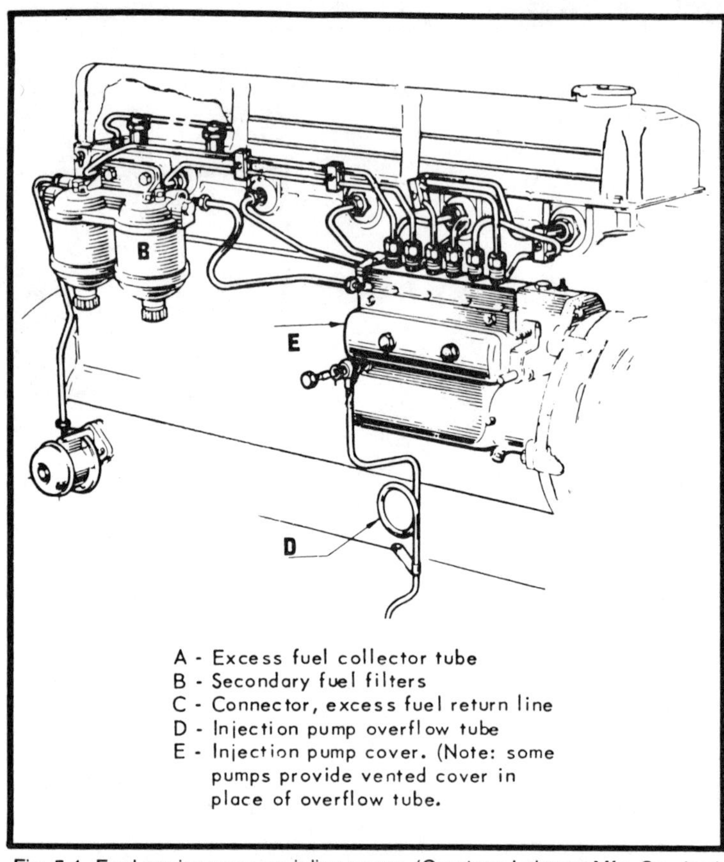

Fig. 5-1. Ford engines use an inline pump. (Courtesy Lehman Mfg. Co., Inc.)

pump by a lift, or low pressure, pump shown in the lower left-hand corner of the drawing. The secondary fuel filters (B) work in conjunction with a primary screen or filter at the tank. The injection pump at E delivers fuel under high pressure to the injectors. Each injector is fed by its own pump plunger. The pump is fitted with an overflow line (D) to return surplus fuel to the tank.

The distributor or "rotary" pump is used by GM Bedford (Fig. 5-2), Perkins, Peugeot, and Waukesha engines, among others. Rather than having individual plungers for each injector, the distributor pump has a single, double-plunger element. The injectors are fed through a rotary valve. Although there are some cost benefits to distributor pumps, the real advantage from a mechanic's point of view is that each injector gets the same amount of fuel.

Individual high pressure pumps are combined with the injectors in Detroit Diesel and Caterpillar engines (Fig. 5-3). The pumps are triggered by a cam-operated rocker arm.

FUEL TANK

The ideal conversion would include a proper diesel fuel tank made of heavy gauge steel and fitted with internal baffles. You can, however, use the existing gasoline tank by making some modifications. The first step is to fit a drain plug if one is not already present. Welding gasoline tanks is dangerous business, so farm this work out to a professional. Remove the evaporative control system. (Figure 5-4 illustrates the system used by many manufacturers.) Discard the charcoal cannister, collector and shield assembly, three-way control valve, and vapor separator. In some cases the vapor return line can be routed into the top of the tank to serve as a fuel return line. Replace the cap with an early-model vented type and add a stage of primary filtration between the tank and lift pump.

LIFT PUMP

The most common lift pump is similar to an automotive fuel pump, but has the advantage of being repairable. Figure 5-5 illustrates one type with a built-in water trap and filter.

German Bosch systems sometimes combine a piston-type lift pump with the injection pump (Fig. 5-6). This device is unusual, in

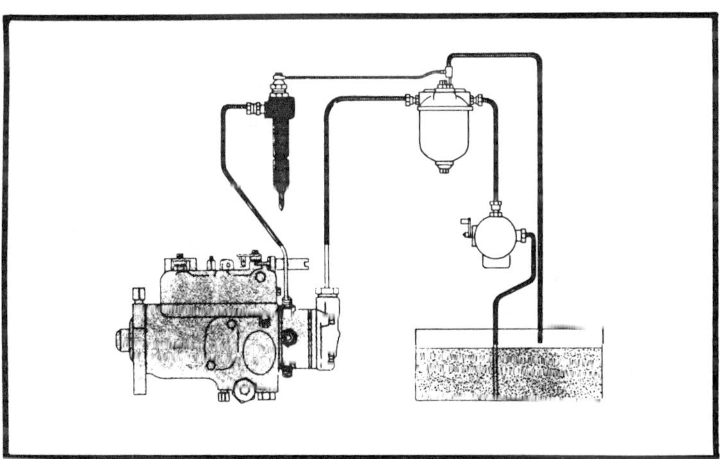

Fig. 5-2. Bedford engines use a distributor pump and have a return lino from each injoctor.

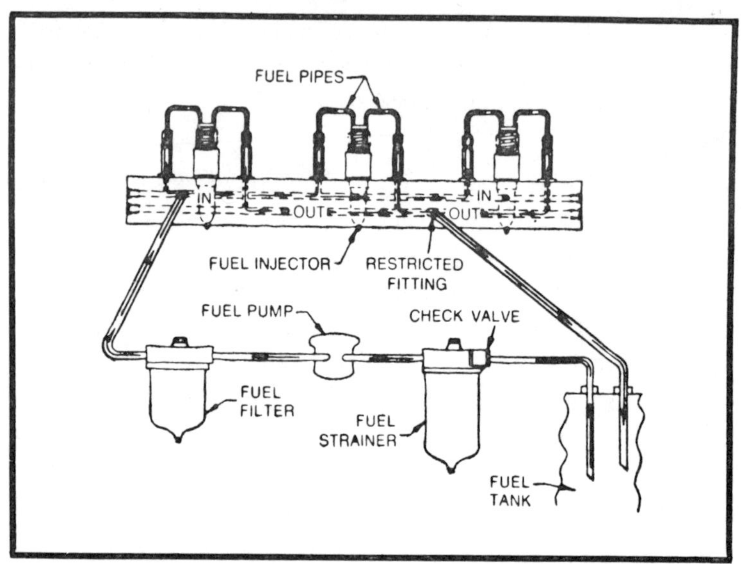

Fig. 5-3. GM's direct injection system as used on Detroit Diesel engines.

having fuel on both sides of the piston, thus eliminating the possibility of vapor lock. Figure 5-7 diagrams the operation. On the downward movement of the piston (View I), the discharge-sidecheck valve (3) closes. Fuel passes through the passage on the left of the piston and

Fig. 5-4. Evaporative emission controls are not needed for diesel fuel and can be discarded.

to the injection pump. The intake side valve (4) is open. In View II the piston is cammed up and fuel enters the outer chamber through check valve (3). Discharge commences in View III.

FILTERS

The fuel must be filtered before it enters the injection pump. Unfiltered fuel, although it may appear clean to the eye, will quickly destroy the pump and injectors. The SAE (Society of Automotive Engineers) recognizes seven classes of oil contamination. On their scale, Class 0 oil is the cleanest, suitable for missile guidance systems and other ultimate-precision technology, while class 6 oil is the

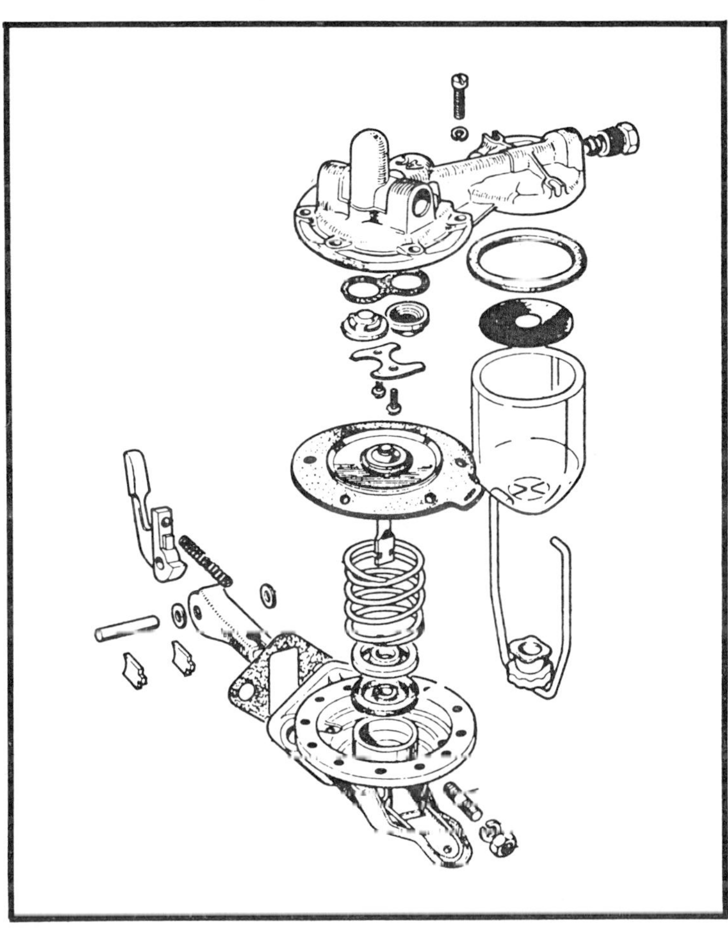

Fig. 5-5. An AC lift pump. (Courtesy GM Bedford Diesel.)

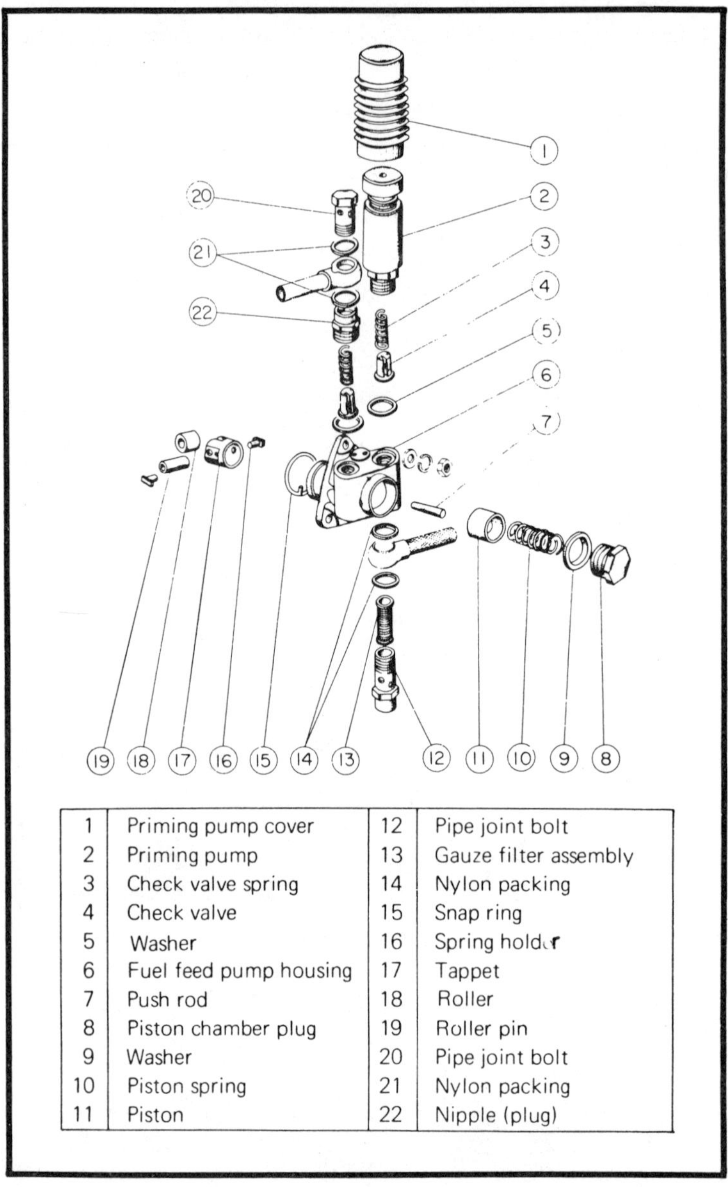

1	Priming pump cover	12	Pipe joint bolt
2	Priming pump	13	Gauze filter assembly
3	Check valve spring	14	Nylon packing
4	Check valve	15	Snap ring
5	Washer	16	Spring holder
6	Fuel feed pump housing	17	Tappet
7	Push rod	18	Roller
8	Piston chamber plug	19	Roller pin
9	Washer	20	Pipe joint bolt
10	Piston spring	21	Nylon packing
11	Piston	22	Nipple (plug)

Fig. 5-6. Exploded view of the lift pump used in conjunction with Bosch injection. (Courtesy Chrysler-Nissan.)

dirtiest. As it comes from the refinery, most diesel oil is class 6 or even lower, in some nether region beyond the SAE classification system. A single cubic centimeter of class 6 oil contains 128,000 particles between 5 and 10 microns in diameter. (A micron is 0.000039 in.). Studies by CAV show that particles in this size-range do most damage to diesel pumps.

Obviously some sort of filter is needed.

Figure 5-8 shows the first line of defense for Bedford engines. The screen keeps out very large particles that would clog the secondary filter, and the bowl traps water. There is no commercially available filter that will stop water; you must have a sediment bowl somewhere in the system. The secondary filter (Fig. 5-9) normally uses a disposable paper element. The vent plug is a convenience when bleeding the system.

Filter elements should be changed often. Some manufacturers suggest changing every 2000 miles. Clean the external surfaces thoroughly before opening the filter case. Clean the inside of the

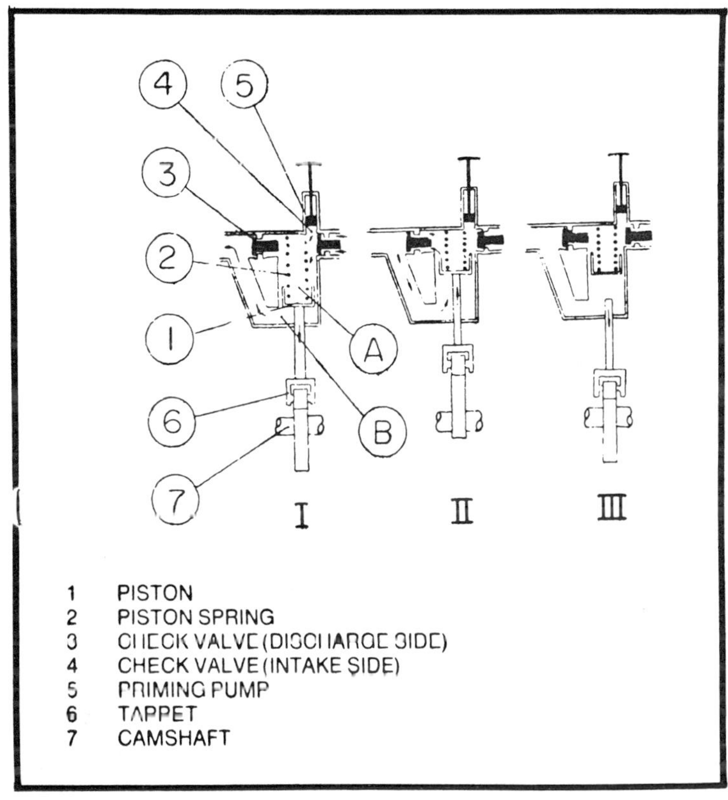

Fig. 5-7. Fuel is present on both sides of the piston. (Courtesy Chrysler-Nissan.)

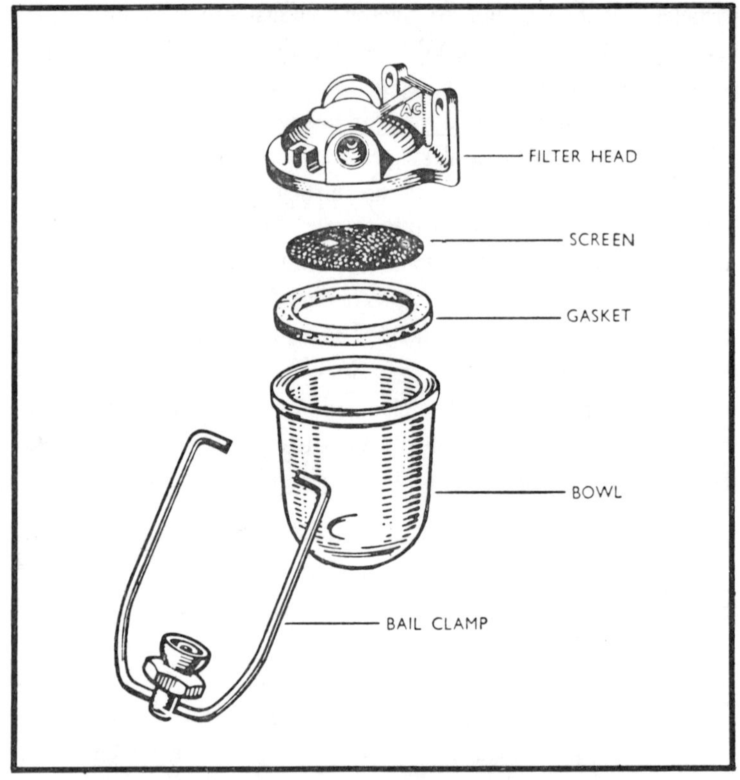

Fig. 5-8. An AC water trap with strainer. (Courtesy GM Bedford Diesel.)

case and inspect the gasket. If the gasket looks at all suspicious, change it along with the element. Bleed the system before attempting to start the engine.

PRIMING PUMP

The priming pump, which is a manually operated pump used to generate pressure for bleeding, is always located in the engine compartment, and is usually part of another component. The Mercedes-Benz "football" pump is on the back of the injection pump; but other firms mount the primer on the lift pump or atop one of the filters.

Air enters the fuel stream through leaks (it is impossible to hermetically seal a fuel system) whenever the system is opened for service or allowed to run dry. A pinched line or a clogged filter will have the same effect.

Air bubbles may lock either or both pumps. The engine may refuse to start, run erratically, or gradually lose speed until it stops. Locate the bleeder screws. You will find them at high points in the system, usually atop the filter cases, and always at the injection pumps (Fig. 5-10).

Follow these steps:

1. Determine that there is enough fuel in the tank. The level must be above the intake pipe.
2. See that the fuel shut-off valve is open.
3. Crack the bleeder screw closest to the tank.
4. Work the priming pump until fuel flows in a steady stream from the bleeder screw.
5. Tighten the screw snugly, but don't overdo it.
6. Repeat the operation on the next bleeder screw, which may be on the outlet side of the secondary filter or on the outboard side of the injection pump.

If bleeding is ineffective, and the system seems to draw in as much air as you pump out, check for loose connections aft of the transfer pump. Sometimes it pays to make a pressure test. Discon-

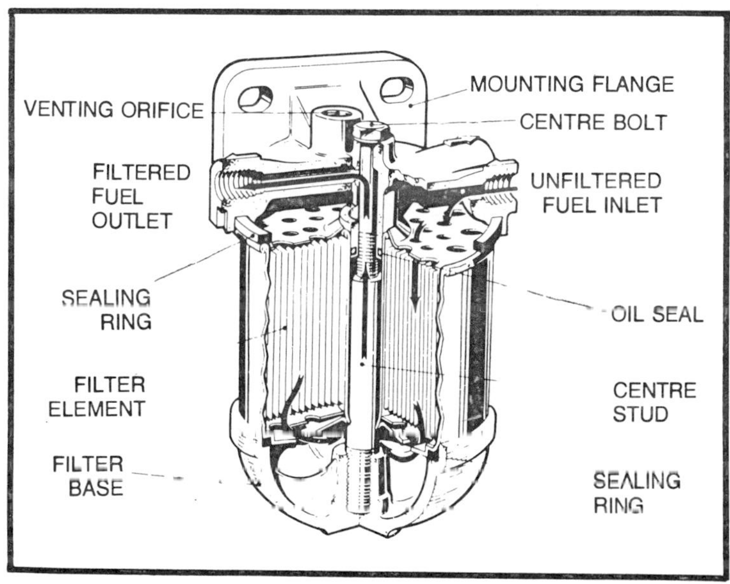

Fig. 5-9. This filter uses the mounting stud as part of the circuit. (Courtesy GM Bedford Diesel.)

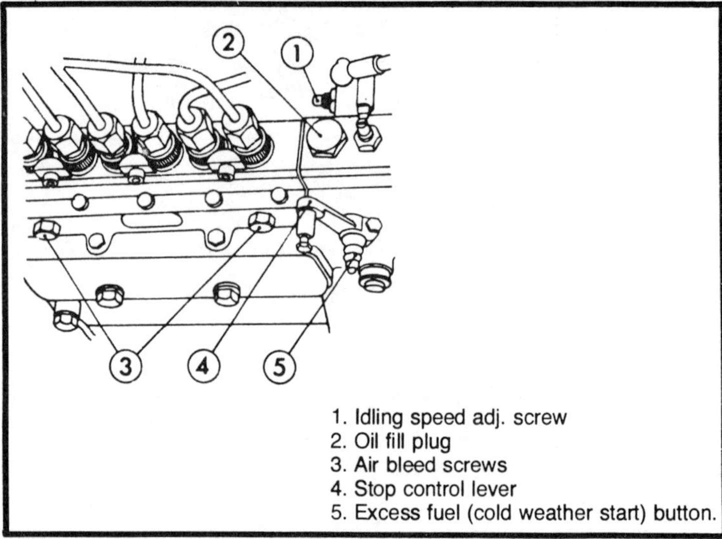

Fig. 5-10. A mechanic will be concerned with the idle speed adjustment screws (1), the cambox oil filler plug (2), air bleed screws (3), stop lever (4), and excess fuel button (5). (Courtesy Lehman Mfg. Co., Inc.)

nect the line on the intake side of the transfer pump and connect it to a regulated and gauged supply of compressed air. Block off the filter cap vent and, if present, the overflow tube. Inspect the cap gasket and tighten the cap down hard.

Introduce very low-pressure air into the system. Industrial and marine tanks can tolerate about 10 psi before they balloon out; but automotive tanks, especially the late-model baffleless types, should not be stressed more than 1 psi. Observe the gauge. Pressure should hold for several minutes. If it doesn't, daub the connections and suspect areas on the tank with soap suds or one of the glycerin-based leak detector solutions used in air conditioning work.

INJECTION PUMPS

Diesel work requires some understanding of the injection pump—if only to appreciate why you shouldn't tinker with it.

Figure 5-11 is a part cutaway of the Bosch type A pump, used on Chrysler-Nissan engines. Early models had the dipstick shown at 58 and very early models had a second dipstick on the governor section. Current production pumps share the oil supply with the engine.

In the operation of the injection pump, there is one plunger (4), for each engine cylinder, so in this case there are six. The plungers ride on roller tappets (14), which in turn are supported by lobes on the camshaft (18). Fuel enters the barrels, or plunger cylinders (3), under transfer pump pressure. As the camshaft turns, the plungers rise and send fuel through the delivery valve (24). The sequence of plunger operation is determined by the firing order of the engine. In the case of Chrysler-Nissan four-cylinder engines it is 1-3-4-2. For the 6-cylinder it is 1-4-2-6-3-5. The pressure developed across the plunger is determined by the pressure required to trip the injectors. The injectors used on these engines lift at 1442.25 psi.

Plunger stroke, which is the absolute distance the plungers move in their barrels—remains the same regardless of engine speed. The effective stroke varies and determines how much fuel is delivered. As you can see in Fig. 5-12, a thread-like groove is milled in the side of each plunger. This helical groove is open to an oil return, or spill, port on the right side of the barrel and is connected to the plunger face by a hole. Turning the plunger changes the relationship between the groove and the oil return port. Under no delivery conditions, as when the engine is ordered to shut down, the groove uncovers the return port as soon as the plunger rises. Output is completely short-circuited and the fuel returns to the tank. Under partial delivery some fuel does get through, since the return port is masked for only part of the plunger stroke tune. At full throttle the groove is turned away from the return port and no fuel escapes injection. Output is then limited only by pump displacement and speed.

The plungers are toothed and meshed with the rack (Fig. 5-13). The rack is moved fore and aft by the throttle and the governor. Fuel delivery valves (part number 1 in Fig. 5-11) mount above each plunger. Their purpose is to chop injection off cleanly, preventing after-dribble. The valve springs (number 20, in Fig. 5-11) have less tension than the injector springs. The valve elements (24) rise off their seats (25) with little reluctance and remain open until the plungers reach their limits of travel. As the plungers descend back down the barrel, some oil is displaced between the plunger and the valve. The spring snaps the valve shut, blocking flow and reducing

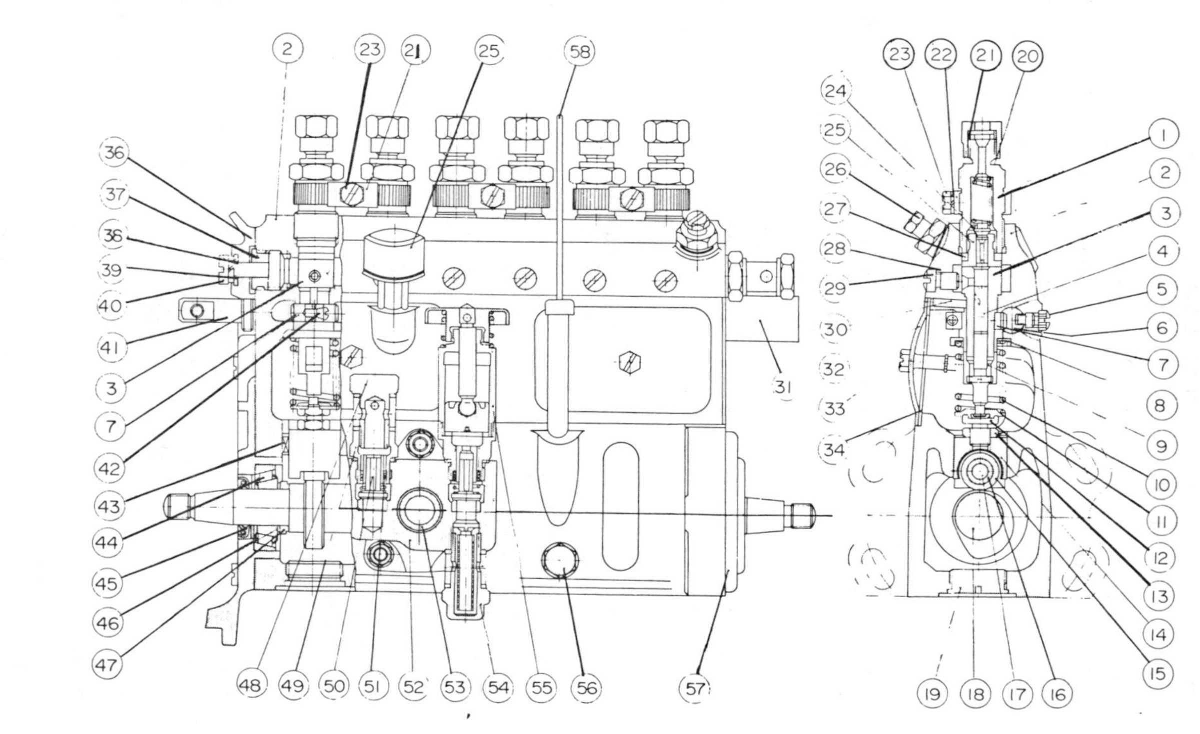

2 23 21 25 58
36 37 38 39 40 41 3 7 42 43 44 45 46 47
48 49 50 51 52 53 54 55 56 57
31
23 22 21 20
24 25 26 27 28 29 30 32 33 34
1 2 3 4 5 6 7 8 9 10 11 12 13 14 15
19 18 17 16

1	Delivery valve holder	16	Roller bushing	31	Cap	46	Adjusting shim
2	Pump housing	17	Roller pin	32	Plate cover	47	Distance ring
3	Barrel	18	Camshaft	33	Cover screw	48	Connector bolt
4	Plunger	19	Screw plug	34	Gasket	49	Nipple
5	Rack guide screw	20	Delivery valve spring	35	Air breather assembly	50	Check valve
6	Control rack	21	Lock plate	36	Governor housing	51	Check valve spring
7	Control pinion	22	Lock washer	37	Screw plug	52	Feed pump housing
8	Upper spring seat	23	Bolt	38	Spring eye	53	Piston
9	Control sleeve	24	Delivery valve	39	Lock washer	54	Connector bolt
10	Plunger spring	25	Delivery valve seat	40	Set screw	55	Priming pump
11	Lower spring seat	26	Air bleeder screw	41	Control rack	56	Drain plug
12	Adjusting bolt	27	Delivery valve gasket	42	Pinion clamp screw	57	Bearing cover
13	Lock nut	28	Gasket	43	Tappet guide	58	Oil level gauge
14	Tappet	29	Screw	44	Tapered roller bearing		
15	Roller	30	Pin	45	Oil seal		

Fig. 5-11. A Bosch type A pump in partial cutaway. (Courtesy Chrysler-Nissan.)

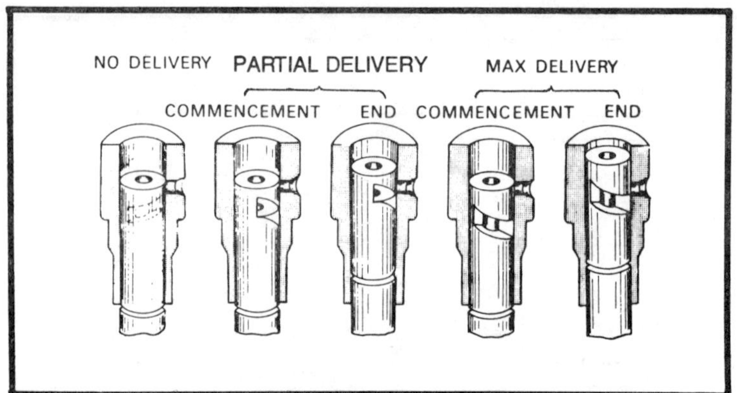

Fig. 5-12. The effective stroke of the plunger depends upon the relationship between its helical groove and the spill port. (Courtesy Chrysler-Nissan.)

pressure in the injection line. The back movement of the valve elements displaces fuel away from the injector.

American Bosch, CAV, and Simms inline pumps follow the same principles as the German Bosch design described here. Their plungers are cam-operated and employ a helical groove for speed control. CAV plungers bleed pressure through vertical slots rather than through drilled holes. Instead of by gears and a gear train, Simms plungers are turned by means of forks that engage the rack.

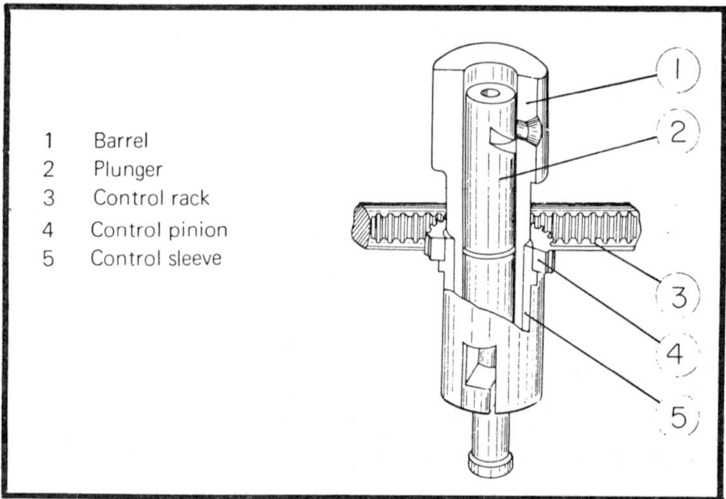

Fig. 5-13. The plunger (2) rotates in the barrel (1) as the rack (3) moves in and out. This motion is transferred by the pinion (4) and the sleeve (5). (Courtesy Chrysler-Nissan.)

CAV Distributor Pump

The CAV model DPA pump is used on several British automotive engines (Fig. 5-14). The pump element consists of two opposed plungers mounted on a rotor. The rotor has as many inlet ports as there are engine cylinders and the stationary hydraulic head has a like number of output ports which are connected to the injectors by high pressure lines.

Figure 5-15 is a somewhat simplified drawing of the pump element, showing the rotor and hydraulic head. During the inlet

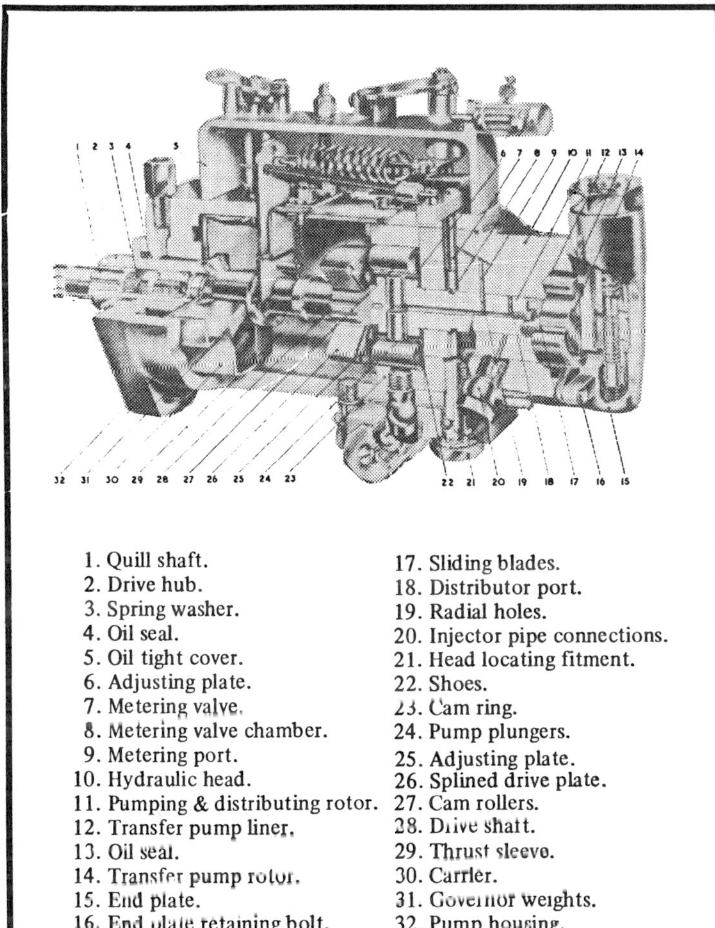

Fig. 5-14. CAV DPA distributor pump.

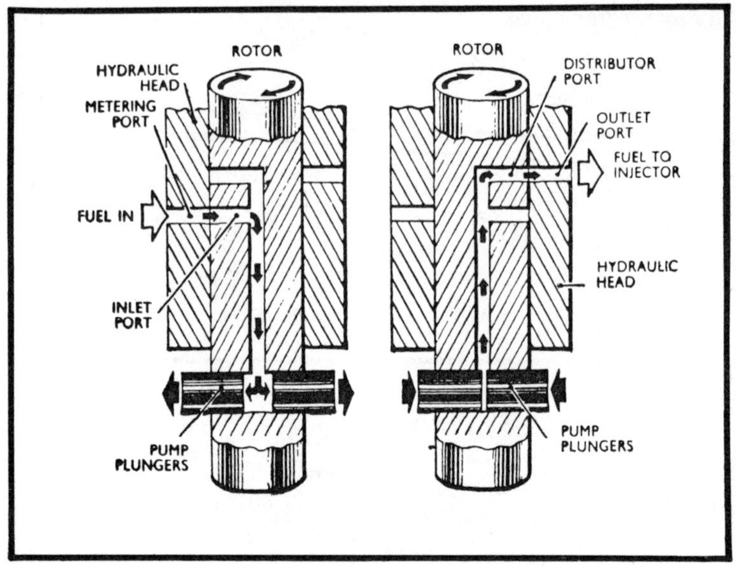

Fig. 5-15. Operation of the DPA pump. The central passage in the rotor indexes with a single fuel inlet, or metering, part and multiple outlet parts.

stroke, fuel enters the rotor through the metering port and passes through the inlet port to the plunger chamber. Fuel pressure forces the plungers apart. As the rotor turns, the metering port is blocked and the distributor port registers with the outlet port in the hydraulic head. At the same time, the plungers make contact with the cam lobes, forcing them together. Fuel leaves the outlet port.

The contour of the cam provides relief near the end of the injection cycle, preventing after-dribble at the injectors. Phasing, which is arranging matters so that fuel delivery begins at the same relative moment for each engine cylinder, depends upon accurately indexing the outlet ports. By the same token, the pump is phased only once, during manufacture. There is no possibility of it getting out of phase in the field.

The distance the plungers move apart on the intake stroke depends upon the amount of fuel that is passed by the metering valve. At idle, relatively little fuel passes and the plungers are only in intermittent contact with the cam. At high speeds or under heavy load, maximum fuel is delivered and the pistons open fully. The amount of fuel the metering valve passes depends upon the positions of the throttle and the governor, as well as upon the speed of the engine-driven transfer pump. At high rpms, transfer pump pressure increases, partially overriding throttle and governor signals.

This system works under four pressures (Fig. 5-16). Low pressure is generated by the lift pump that delivers fuel to the transfer pump. Transfer pump pressure is somewhat higher, and, as indicated above, tends to rise with engine speed. The regulating valve imposes a limit on transfer pressure by blocking part of the fuel input. Fuel in the metering valve is at less than transfer pressure because of flow resistance through the port. The diameter of this orifice varies with the application. Injection pressure is determined by the force needed to unseat the injectors. A typical figure is 200 atmospheres (2940 psi).

Roosa Master

The Roosa Master model DB pump is found on several American engines, including the popular Waukesha VR series. Since Roosa

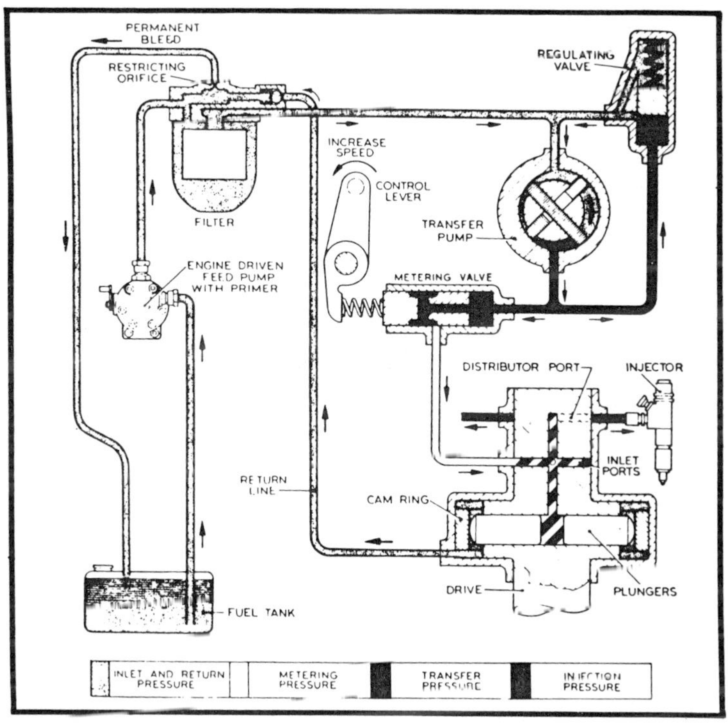

Fig. 5-16. The fuel circuit on Bedford engines.

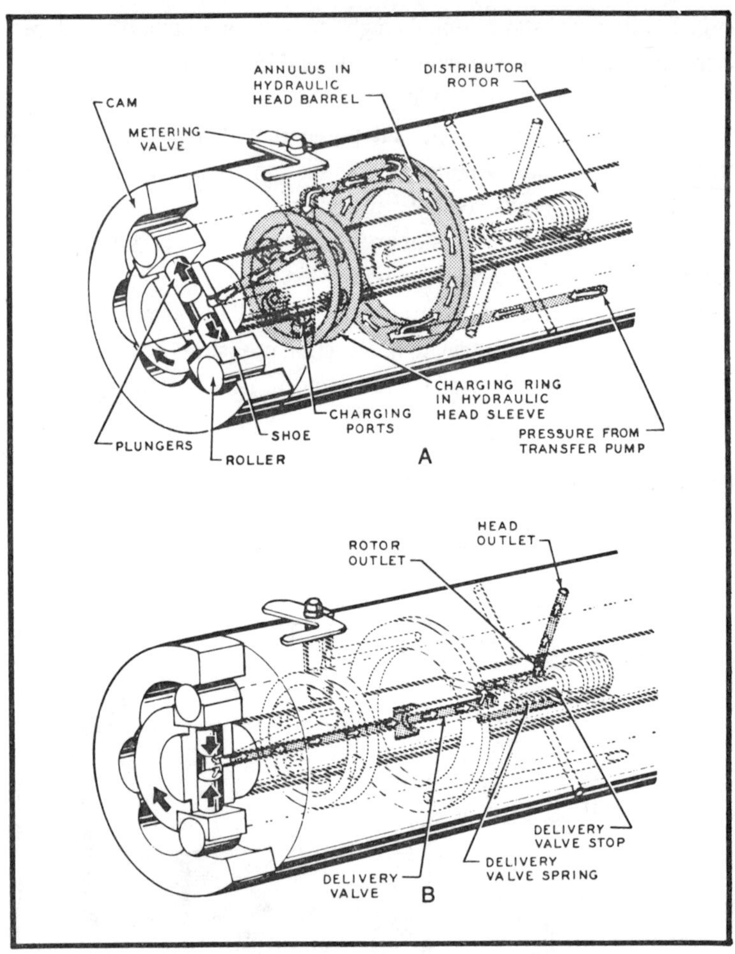

Fig. 5-17. Operation of the Roosa Master DB pump. During charging fuel flows through the annulus and one charging port (view A). Continued rotation cams the pistons together to discharge the fuel past the delivery valve and to the appropriate head outlet port (view B). (Courtesy Waukesha Motor Co.)

Master holds the basic patents, this pump is similar to the license-built DPA.

The transfer pump sends fuel through the inlet port in the hydraulic head, into a circular groove, or annulus, and through the metering valve (Fig. 5-17A). From there the fuel passes to the charging ring which opens to the charging ports (Fig. 5-17B). There are as many charging ports as there are cylinders. The charging ports empty into the space between the plungers.

The amount of fuel between the plungers determines how far they will move. At low speeds very little fuel is admitted, while at high speed or under heavy loads the plunger movement is only limited by the leaf spring adjustment.

As the rotor turns, the charging port is blocked. Hydraulically, things are static until the rotor brings the discharge port into register with the outlet port (Fig. 5-17B) Both rollers contact lobes on the cam, and the plungers move together. The fuel between them moves into one of the injectors.

The delivery valve is located in the center of the rotor, athwart the discharge port (Fig. 5-18). Unlike delivery valves used on inline pumps, this valve does not come to rest on a seat. Sealing is provided by the fit between the valve plunger and the bore. When injection begins, fuel pressure moves the valve plunger out of its bore and opens the displacement section (A). Fuel enters the spring chamber which has been made deliberately large for reasons that should become clear in a moment.

At the end of injection, the cam releases the plungers. Pressure on the pump side of the valve drops, and the valve spring forces the

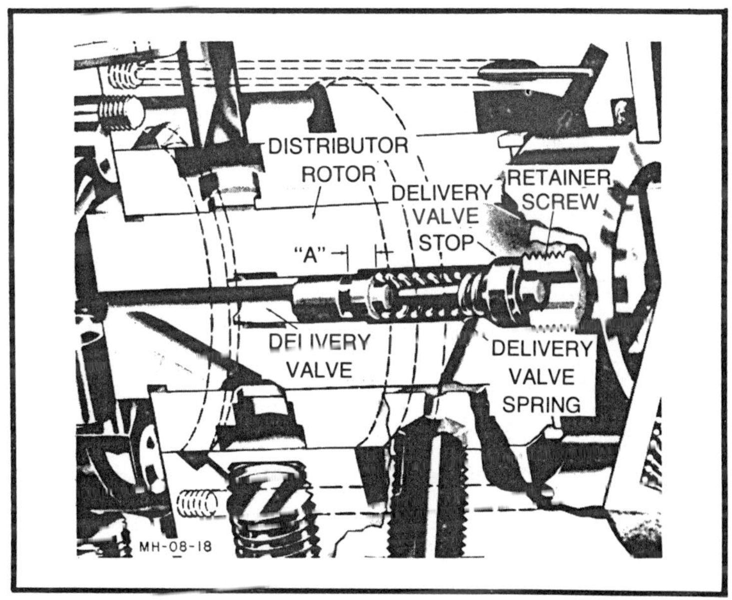

Fig. 5-18. The Roosa Master delivery valve in cutaway view. (Courtesy Waukesha Motor Co.)

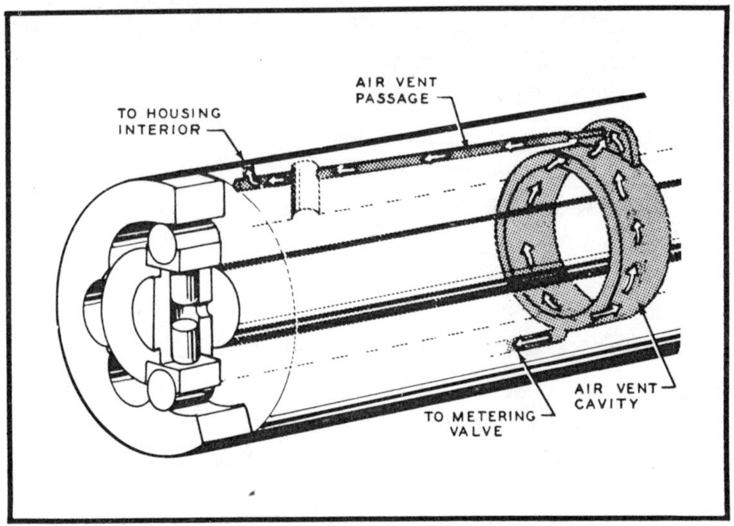

Fig. 5-19. The air vent cavity and passage automatically bleeds air from the system. (Courtesy Waukesha Motor Co.)

valve back toward the pump. The retreating valve increases the displacement of the spring cavity which is no longer full of fuel. Meanwhile, the discharge port is still in register and open to the injector. Fuel in the injector line responds to the pressure loss in the spring cavity and flows back through the discharge port to equalize pressure. This whole mechanism—of valve, spring cavity, and log—in closing the discharge ports creates a sudden loss of line pressure. The injector shuts off immediately without fuel-wasting dribble.

Another interesting feature is the return oil circuit (Fig. 5-19). Fuel enters the pump by way of an air vent cavity. Air in the fuel is diverted up through the cavity and to the air vent passage. Since some fuel may be borne along with the air, the vent is connected to a tank return line.

MAINTENANCE AND REPAIR

The most important part of pump maintenance is to change the fuel filter elements at prescribed intervals. Some diesel builders substitute after-market filters for those originally supplied with the engine to provide greater capacity and finer filtration. Cambox oil sumps should be checked almost daily and drained every 2000 miles.

Timing. All pumps have some sort of adjustable drive to allow adjustment of the timing and the ignition point in the field. Figure 5-20 illustrates the marks for the Roosa Master pump, which align above the center of the window on this particular model. Other Roosa pumps have the marks aligned below or at the center of the window. Number 1 cylinder should be on its compression stroke when the piston is a specified number of degrees before top dead center. The vibration damper or the flywheel is usually referenced. If it is not referenced, you can obtain a roll of degree-tape from the manufacturer. The 0 mark on the tape corresponds to top dead center.

Timing drills vary between manufacturers. The drill for the Amercian Bosch PSU series pump is instructive since it makes allowance for all the variables that can affect pump timing.

As fitted to current Onan engines, delivery begins 19 crankshaft degrees before top dead center, although earlier engines had more advance with 21 degrees. Top dead center and delivery marks are stamped on the flywheel (Fig. 5-21).

Remove the pump from the engine and install a standard timing button under the plunger tappet (Fig. 5-22). The timing button is a spacer that raises or lowers the pump plunger in its barrel. The standard button is either unmarked or stamped No. 11. Turn the flywheel to bring the PC mark into alignment with the pointer on the compression stroke for No. 1 cylinder. PC stands for "port closed," referring to the spill port. Delivery commences as soon as the spill port is closed.

Remove the timing screw, located on the pump body just above the mounting flange. Insert a 0.125 in. brass wire into the timing screw hole (Fig. 5-23). Turn the drive gear until the wire bottoms. This is the signal that the spill port is closed.

Mount the pump on the engine, making sure that the shims are in place and that the flange O-ring is in good condition. Withdraw the brass wire and replace the timing screw.

The pump will now be in rough time since its delivery point will correspond with the PC mark on the flywheel, located 19 or 21 degrees before top dead center. The important matter of the button thickness remains. The "standard" button is one out of 16 possible

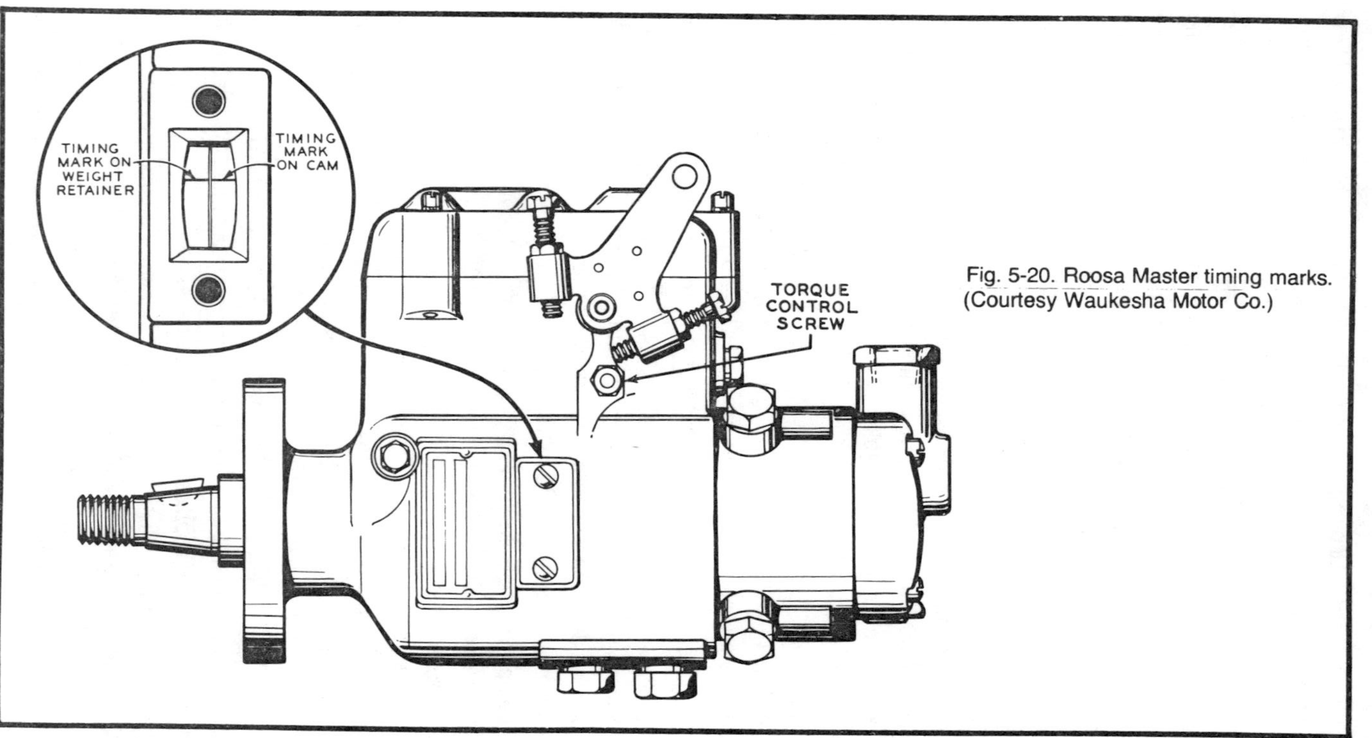

Fig. 5-20. Roosa Master timing marks. (Courtesy Waukesha Motor Co.)

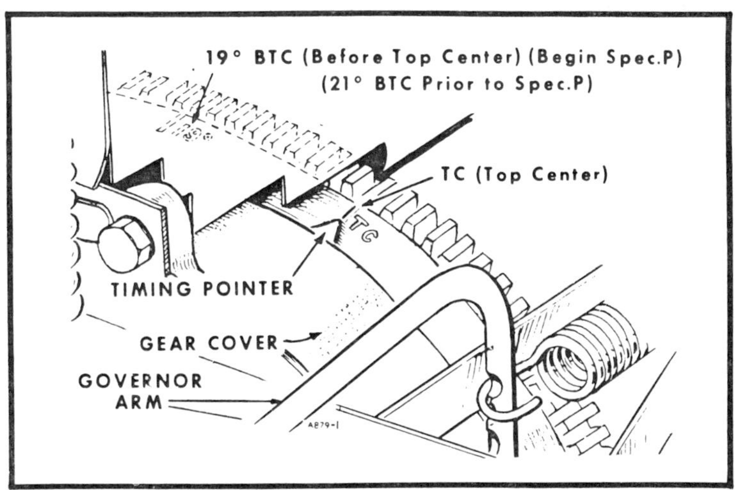

Fig. 5-21. Top dead center is marked on the flywheel or on the vibration damper. Turning the flywheel a few degrees against engine rotation will disclose a second mark that signals the beginning of injection. (Courtesy Onan.)

choices. The odds are 15 to 1 that your button will be the wrong thickness and that delivery will be early or late.

The pump must be flowed to visually determine when the fuel spill port closes. Remove the delivery valve spring (Fig. 5-24). This disables the valve so that there is no resistance to flow through the pump. Turn the flywheel to about 15 degrees before the PC mark with No. 1 cylinder on its compression stroke. Set the fuel control to full delivery. Disconnect No. 1 injection line at the nozzle and position the line so that flow is visible. Pump the priming lever while

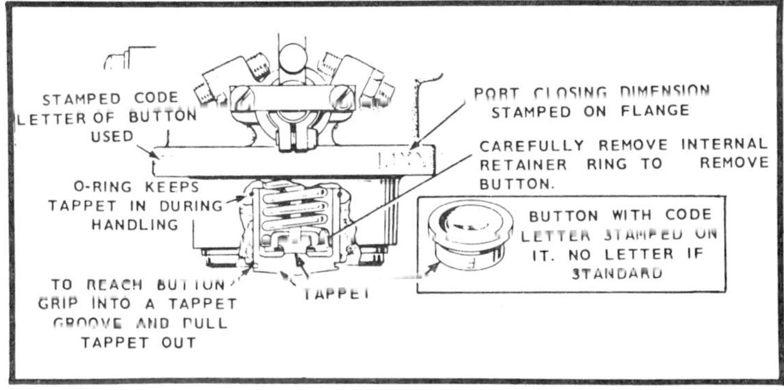

Fig. 5-22. American Bosch series PSU pumps are finally timed by a button under the plunger tappet. (Courtesy Onan.)

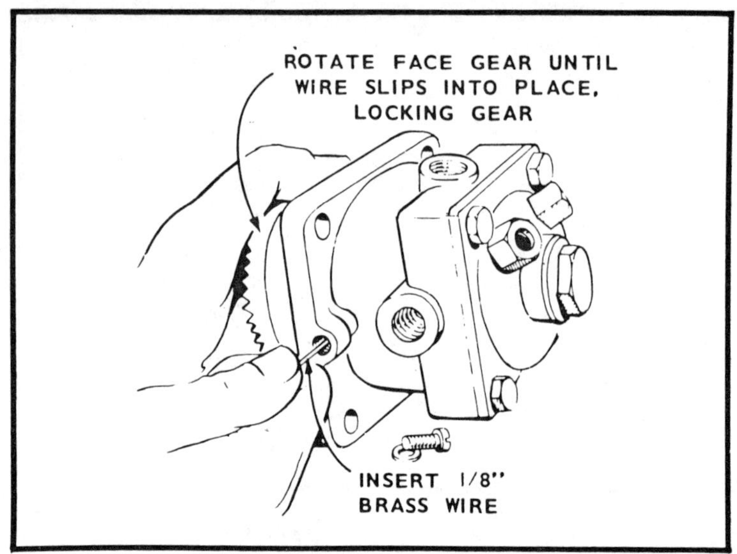

Fig. 5-23. The delivery valve must be disabled or removed before the pump can be timed by flow. (Courtesy Onan.)

turning the flywheel toward the PC mark. The position of the flywheel when fuel flow stops is the "port closing point," the point where injection begins. If your button is the correct thickness, delivery should stop at the PC mark. If delivery is early, use a thinner button; if late, use a thicker one.

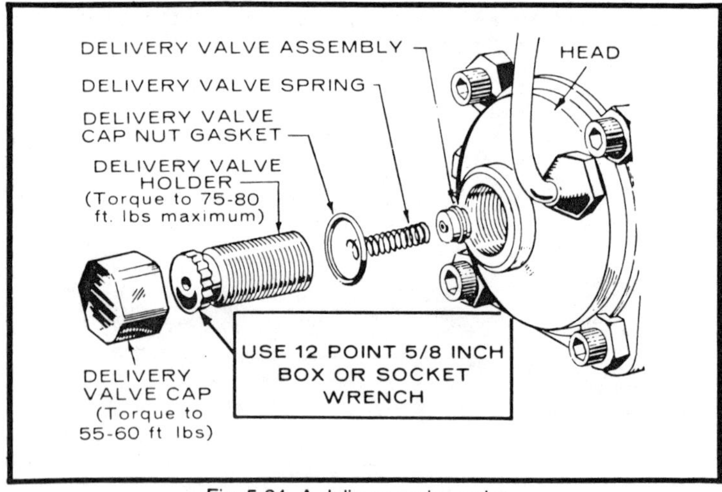

Fig. 5-24. A delivery spring valve.

You may wonder why fuel flow stops at the port closing point. Fuel enters the plunger cavity by way of the spill port when the plunger is low in the barrel. Fuel escapes through the same spill port, through the helical groove in the side of the plunger. During the timing drill, the pump plunger is moved very slowly and cannot develop enough back pressure to overcome the pressure generated by the priming pump. Consequently, fuel flows through the spill port and into the barrel chamber. It continues to flow until the spill port is no longer aligned with the helical groove. If the pump were running at speed, the injection would commence when the port closed.

Onan suggests that fuel be monitored at the injection line. While this method is adequate, greater accuracy can be obtained by replacing the line with a swan-neck fitting (Fig. 5-25). The chamber on the end of the pipe gives a very close indication of fuel shutoff.

A timing light may not sound very appropriate for diesel engines, but there are models on which it can be used. Figure 5-26 illustrates the hookup for the Sun light. An injection line expands under pressure. The light is triggered by a transducer fitted over No. 1 line, usually just above the delivery valve. Power for the Xenon bulb is from the vehicle's 12V or 24V battery bank or from a 115 VAC receptacle. The engine is run at a slow idle and the stroke is pointed at the crankshaft timing marks. A dial on back of the tool is calibrated in revolutions per minute and shows the number of de-

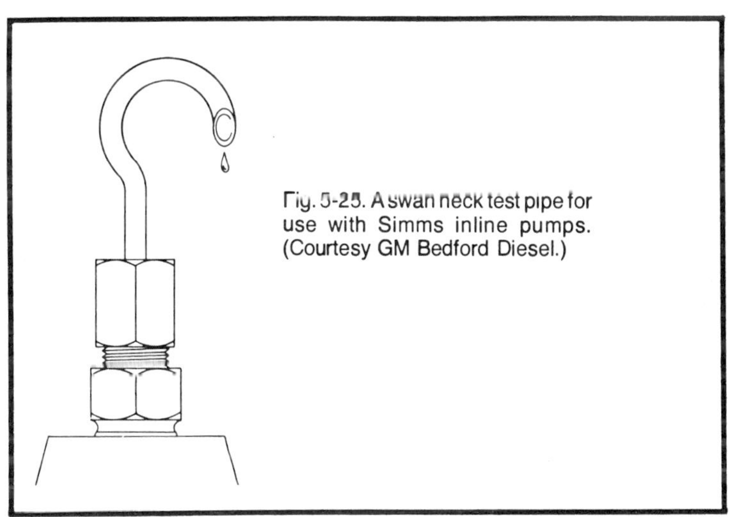

Fig. 5-25. A swan neck test pipe for use with Simms inline pumps. (Courtesy GM Bedford Diesel.)

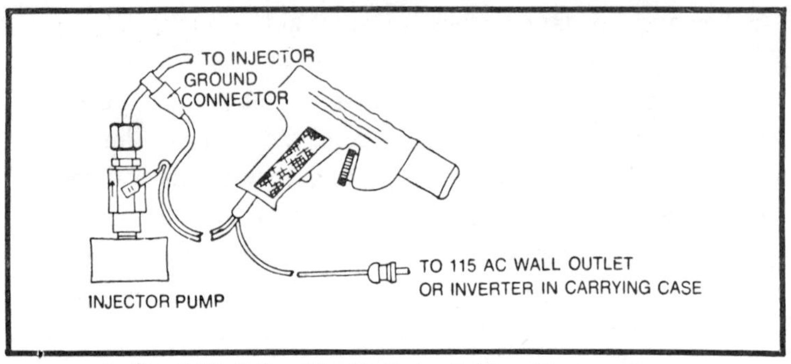

Fig. 5-26. A Sun Electric timing light and tachometer.

grees of pump advance before TDC. The engine can be accelerated to check the automatic advance as explained below.

Phasing and Calibration. Serious pump work can be a specialty in its own right, requiring special equipment and training (Fig. 5-27). "Phasing" refers to the internal timing of the plungers in an inline pump. For a four-cylinder engine, the plungers should deliver fuel in the sequence set by the firing order and 90 degrees of camshaft rotation (or 180 degrees of crankshaft rotation) apart. A six-cylinder should fire every 60 degrees. Phasing also includes measuring vertical plunger travel and head clearance, or the distance from the top of the plunger to the stop on the roof of the plunger chamber.

"Calibration" is the tailoring of pump output to engine and governor specifications. These requirements are plotted against pump rpm and control rack travel. The output is collected, then measured after a predetermined number of strokes. Allowable variation is 2.5% or less, depending upon the engine manufacturer.

Figure 5-28 illustrates a Bacharach Model U-7500A pump calibration stand. It is powered by a 7.5 hp motor with speed ranges from 45 to 4500 rpm. The instrument has its own supply of temperature-controlled calibration oil. A 10°F change in oil temperature can throw delivery off as much as 1.5%.

HIGH PRESSURE LINES

The high pressure fuel lines or pipes connect the jerk pump with the injectors. They are made of thick section steel tubing and fitted with union nuts at each end.

Exercise extreme care when handling these. Clean the external surfaces thoroughly and detach the steady bracket before opening a line. Do not bend a line for clearance when changing an injector or as an expedient when testing an injector out of the engine. Cap the lines, the fuel delivery valves, and the injectors as soon as they are open to the atmosphere. Before assembly, both new and used lines, however carefully capped, should be cleaned in solvent, blown dry with filtered compressed air, and sunk in fuel.

These lines are subject to pressures in the neighborhood of 200 atmospheres. Small imperfections act as stress risers, particularly where the union nut threads over the flare. Figure 5-29 shows the effect.

Check the seating surface on the delivery valve holders and injectors. The flare will leave a "footprint," showing where it stood during assembly. Overtorquing the union will send the flare off-center. Once this happens both valve and injector fittings should be discarded since they exert a bending force on the lines.

INJECTORS

The injectors are the final and in many respects the most critical components of the fuel system. The injectors must:

- Contain engine compression.
- Divide raw fuel into a fine spray for combustion.
- Direct the spray pattern deeply into the chamber, but not against the piston top or the cylinder walls. Should this happen, the fuel will be difficult to ignite and, once ignited, might burn through the engine metal.
- Stop injection promptly.

Injectors are classified by the design of the nozzle and by the location of the pump.

An open-hole nozzle is just that: an open hole in the nozzle tip through which fuel enters the cylinder. A check valve in the injector body triggers delivery and prevents compression leak back into the fuel line. While admirably simple, open-hole injectors are impractical for automotive engines.

Automotive injectors have "closed" nozzles in which the nozzle orifice is blocked by a differential valve between injection periods.

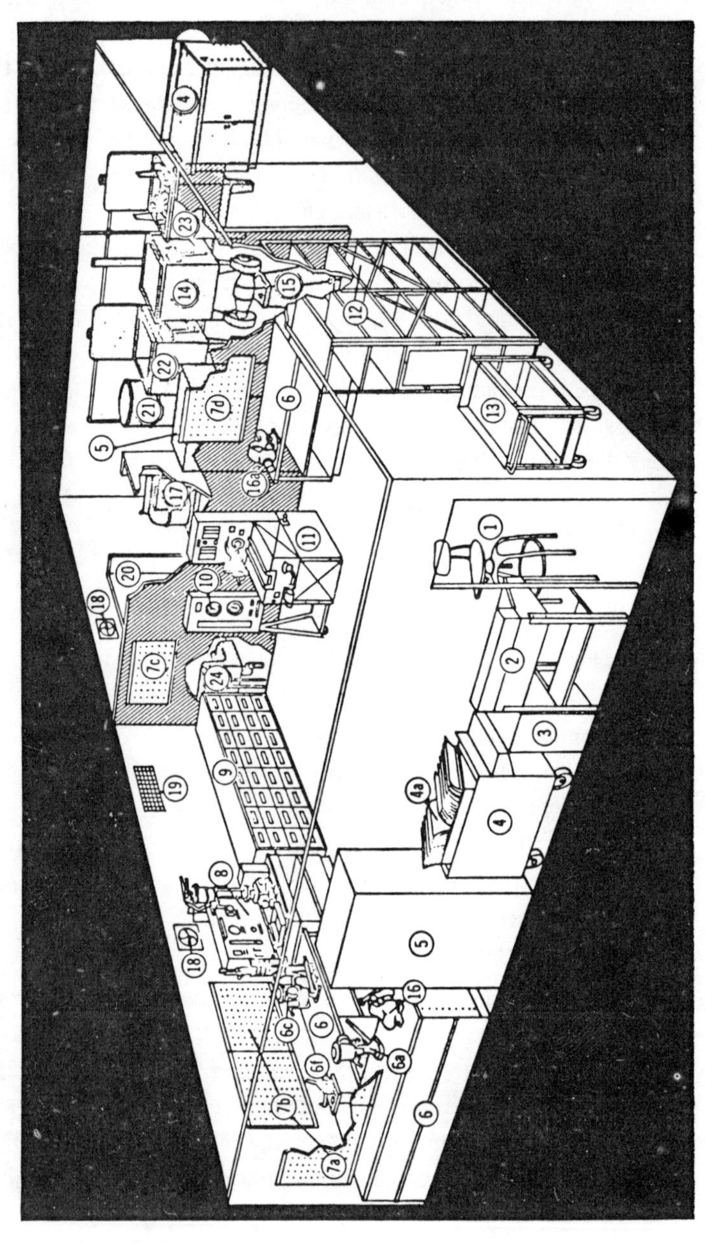
1
2
3
4
4a
5
6
6a
6c
6f
7a
7b
7c
7d
8
9
10
11
12
13
14
15
16
16a
17
18
19
20
21
22
23
24

1. Stool
2. Desk
3. Filing cabinet
4. Portable work bench
4a. Service manual rack
5. Storage cabinet
6. Work bench (stationary)
6a. Nozzle test pump with dial gauge (65-0030)
6c. DDED injector tester (65-1019)
6f. PT injector cup assembly fixture (66-7017)
7. Pegboard
7a. Nozzle servicing tools
7b. Injector servicing tools
7c. High pressure fuel injection pump lines
7d. Pump and governor servicing tools
8. Injector flow comparator (67-7057)
9. Storage bins
10. Flow test panel (67-7249)
11. Diesel fuel injection pump test stand U-7500A (67-7018)
12. Shelves
13. Steel cart
14. Cleaning machine
15. Grinder-buffer on stand
16. Industrial vise
16a. Pump mounting vise
17. Standard 5 h.p. air compressor with tank
18. Air exhaust system
19. Air conditioning inlet system
20. Spray paint booth
21. Test oil drum reserve
22. Leakage test tank
23. Cleaning tank
24. Sink

Fig. 5-27. The well equipped diesel pump and injector facility should look like this. (Courtesy Bacharach Instrument Company, division of Ambac Industries, Inc.)

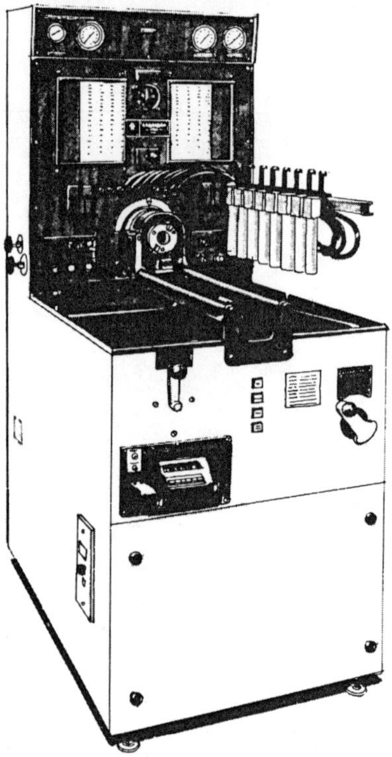

Fig. 5-28. A Bacharach pump calibration stand.

Figure 5-30 is a cross-section of the Peugeot-supplied DN OSD 230 injector, and the next drawing, Fig. 5-31, is an exploded view of a German Bosch injector used in Ford engines. The upper part, or stem, of the injector valve bears against a coil spring. The shoulder

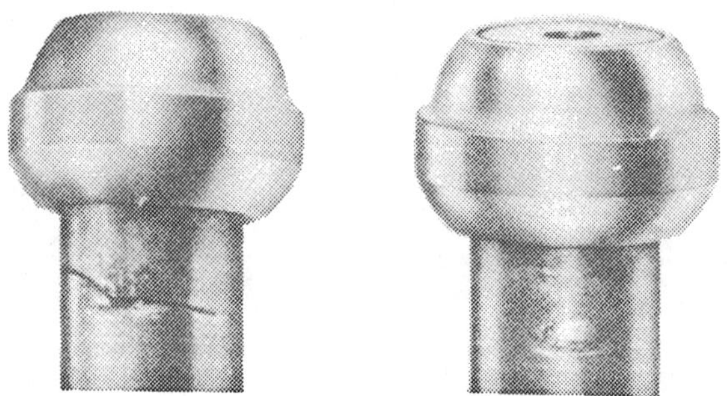

Fig. 5-29. Faulty injection lines. (Courtesy GM-Bedford Diesel.)

on the lower part of the valve is subject to injection pump pressure. The tip of the valve forms the seat.

For injection, fuel passes through a channel along the side of the injector body and collects at the tip of the valve. The collection point is known as the cup, or pressure chamber (Fig. 5-32). Fuel is brought down low in the injector because it is needed as a lubricant and because the heat it picks up from the engine metal helps to atomize it. When pump pressure overcomes valve spring tension, the injector fires. Once the pump plunger passes the limit of its stroke, pressure falls. The delivery valve shifts toward the pump and the valve spring closes the injector valve against its seat.

The units shown in Figures 5-30, 5-31, and 5-32 are pintle injectors, the type most often found in automobile engines. The spray pattern is narrow and long-ranging, making this type of injector the first choice for precombustion chambers. The injector gets its name from the extension or "pintle" on the valve tip. The pintle helps to direct the spray, but its more important function is to limit the rate of fuel delivery (Fig. 5-33). The pintle remains in the nozzle orifice, restricting flow, until the valve is full open. The throttling

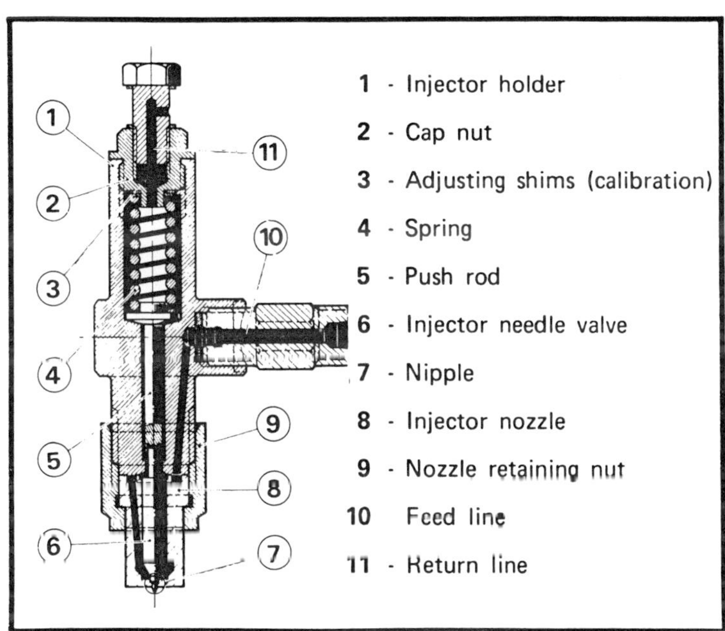

Fig. 5-30. Cutaway of a Bosch injector used by Peugeot.

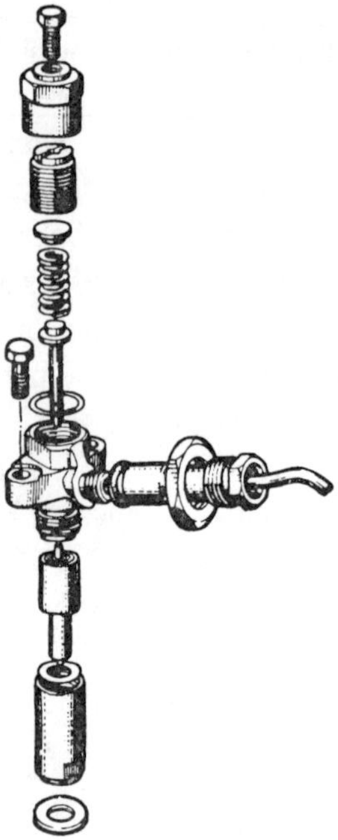

Fig. 5-31. Exploded view of a pintle injector. (Courtesy Lehman Ford Diesel.)

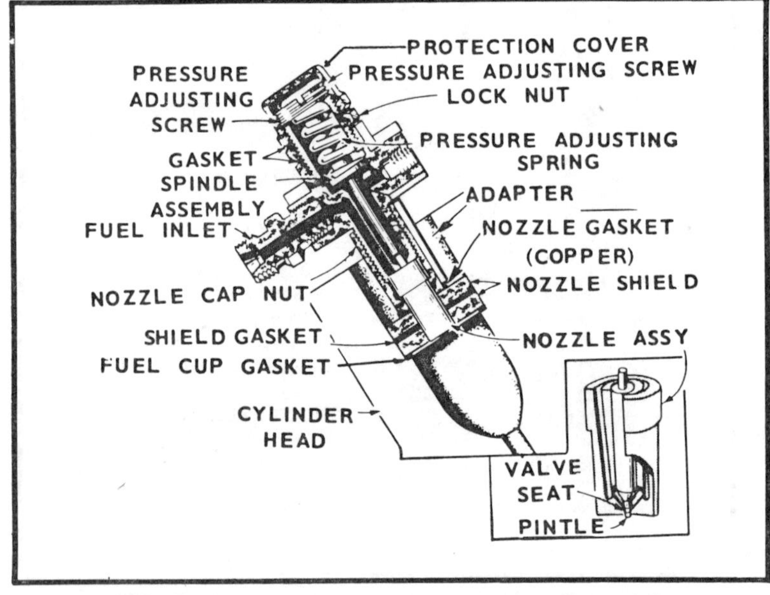

Fig. 5-32. Cutaway of an American Bosch injector. (Courtesy Onan.)

effect of the pintle is most noticeable at idle. When less fuel is admitted during any given time period, cylinder pressures rise more slowly and there is less "diesel clatter."

Unit injectors combine the pump and injector body in the same assembly. The pump plunger is driven by a rocker arm whose free end rides on an engine camshaft. The high pressure fuel circuit is confined within the injector body, and is less likely to leak than if the system had involved a remote pump and fuel lines.

Multihole injectors have several fuel orifices, arranged symmetrically around the tip. Some have as many as 18 orifices, 0.006 in. in diameter. The spray pattern is usually symmetrical, although Mack is an exception. Because of valve interference, Mack injectors are mounted at an angle in the chamber. The spray pattern is offset a few degrees to compensate.

Standing waves and the phenomenon of "phantom" injection also offer an advantage. When the injector opens, a low pressure wave moves through the fuel column to the pump. A high pressure wave is reflected back when the fuel encounters the injector nozzle. These waves move at 5000 feet per second, and, depending upon shape and complexity of the fuel lines, can trigger early injection or delay normal injection. Mounting the pump in unit simplifies the wave pattern.

The injector section works like the pintle injector already described, and shares the same general parts arrangement (Fig. 5-34).

The pump section is reminiscent of a single cylinder from an inline pump (Fig. 5-35). The plunger is triggered by a cam follower and its effective stroke is controlled by a rack. As the rack moves,

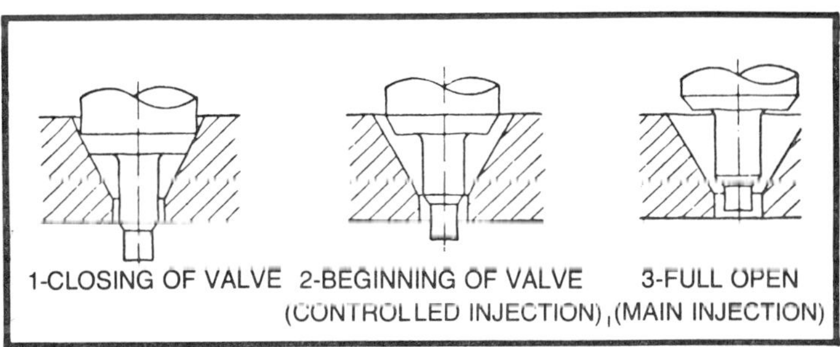

Fig. 5-33. The throttling effect of the pintle slows initial fuel delivery. (Courtesy Chrysler-Datsun.)

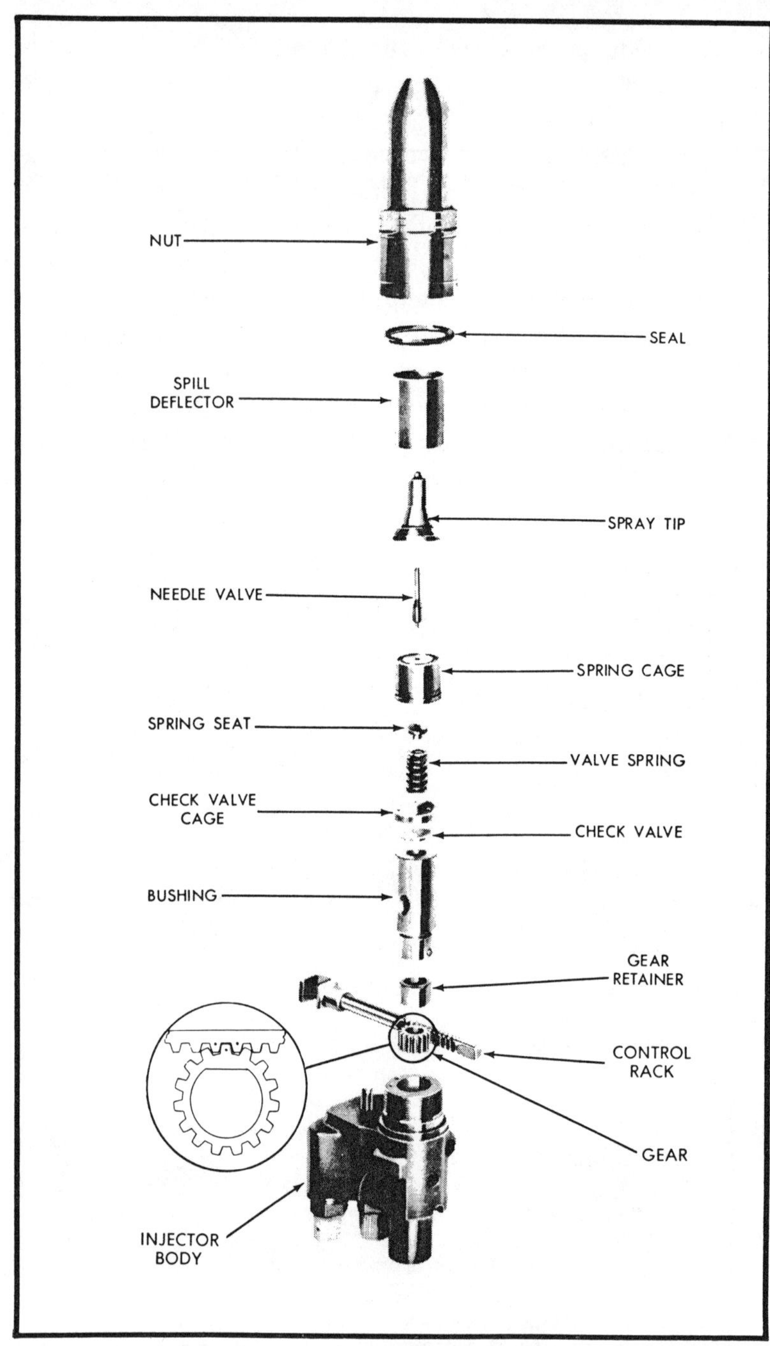

Fig. 5-34. A Detroit Diesel injector section.

the plunger turns and progressively exposes a helical groove on the plunger body to a spill port. Fuel not injected, along with fuel circulated for lubrication and cooling, is returned to the tank.

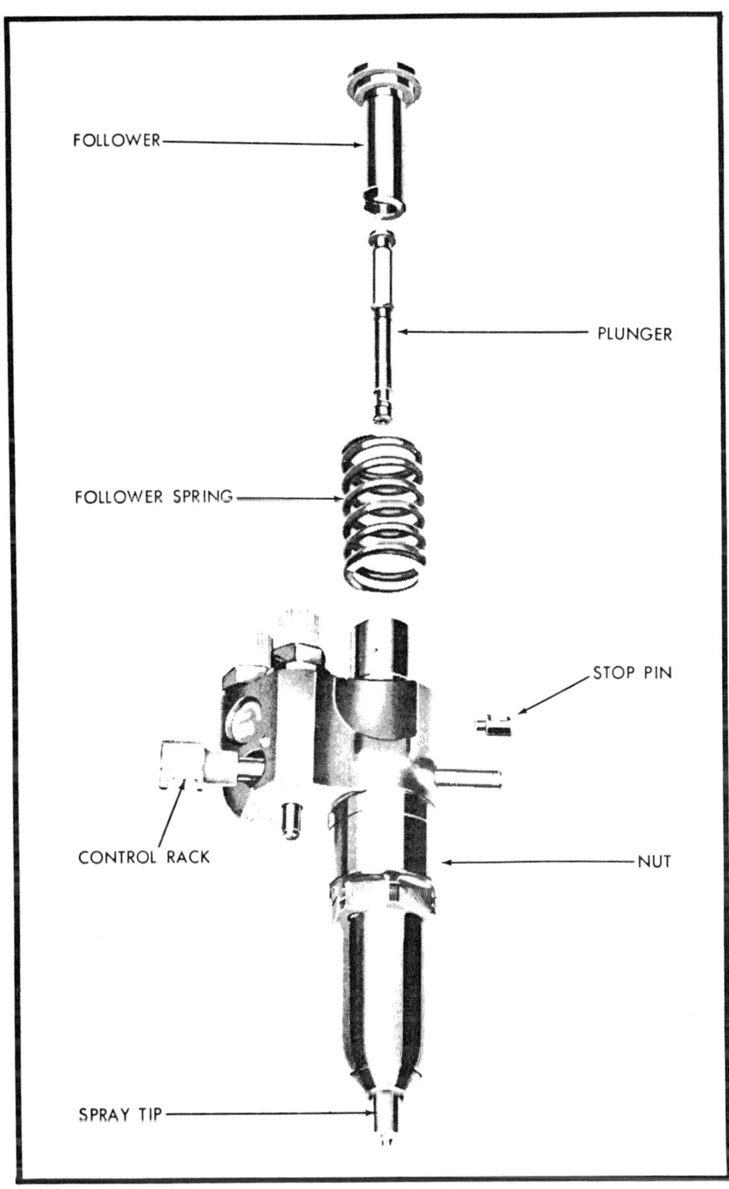

Fig. 5-35. A Detroit Diesel pump section.

Fig. 5-36. Some injectors must be pried loose with a special tool. (Courtesy Ford Engine & Turbine Operations.)

Service

Injectors, like spark plugs, are to be assumed guilty until proven innocent. A faulty injector may give one or more of the following symptoms:

- Black or gray smoke: If the smoke issues in puffs, the injector or pump section for a single cylinder is involved.
- White smoke: Evidence that one or more cylinders are misfiring.
- Detonation: Caused by early injection or after-dribble.
- Uneven running, erratic idle, loss of power.

Bleed the system first and then isolate the affected cylinders.

Identifying which injector is at fault is not difficult. Operate the engine at idle. Disable one injector at a time and listen for an rpm drop. The injector or injectors that make no, or very little, difference in engine speed will be at fault. Unit injectors can be disabled by depressing the pump follower with a screwdriver. Jerk pump injectors can be taken out of the circuit by cracking either of the union nuts on the high pressure line.

To change an injector, remove the fuel line bracket and disconnect the affected line at both ends. Back the union nuts out evenly so the line "floats" between the nuts and is not subject to twisting or bending forces. Cap the ends of the line, the delivery valve, and the injector fitting.

The standard mounting arrangement is two bolts through a flange on the injector body. The flange rests on a machined surface on the head and is gasketed with a copper ring. Simms injectors must be pried loose after the nozzle body bolts are out. A special tool is available for this purpose (Fig. 5-36).

Clean the injector sleeve with a small wire brush. Coat the sleeve with grease to prevent carbon from falling into the cylinder. If the sleeve is cracked, it may have to be repaired. However, check with a specialist first: some engines are more tolerant of sleeve cracks than others. The sleeve shown in Fig. 5-37 is acceptable in Peugeot engines.

The injector should be tested for opening pressure, seat and valve-stem leakage, chatter, and correct spray pattern. A test pump will be able to perform all of these tests, although the spray pattern is sometimes checked by temporarily connecting the injector to the pump and cranking the engine (Fig. 5-38). Be careful to keep eyes and hands clear of the spray, since it will penetrate tissue, causing blindness or blood stream infection. Figure 5-39 shows the desired spray pattern for one popular pintle-type injector.

Install the injector over a new copper gasket, unless the gasket is integral with the head. Remove the caps.

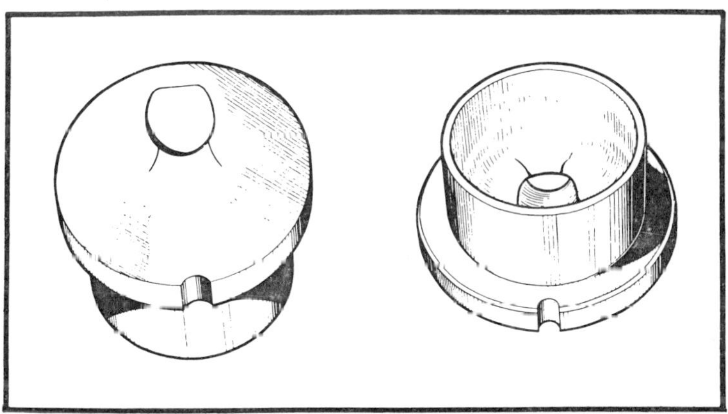

Fig. 5-37. Some engines are tolerant of small cracks in the injector sleeve. (Courtesy Peugeot.)

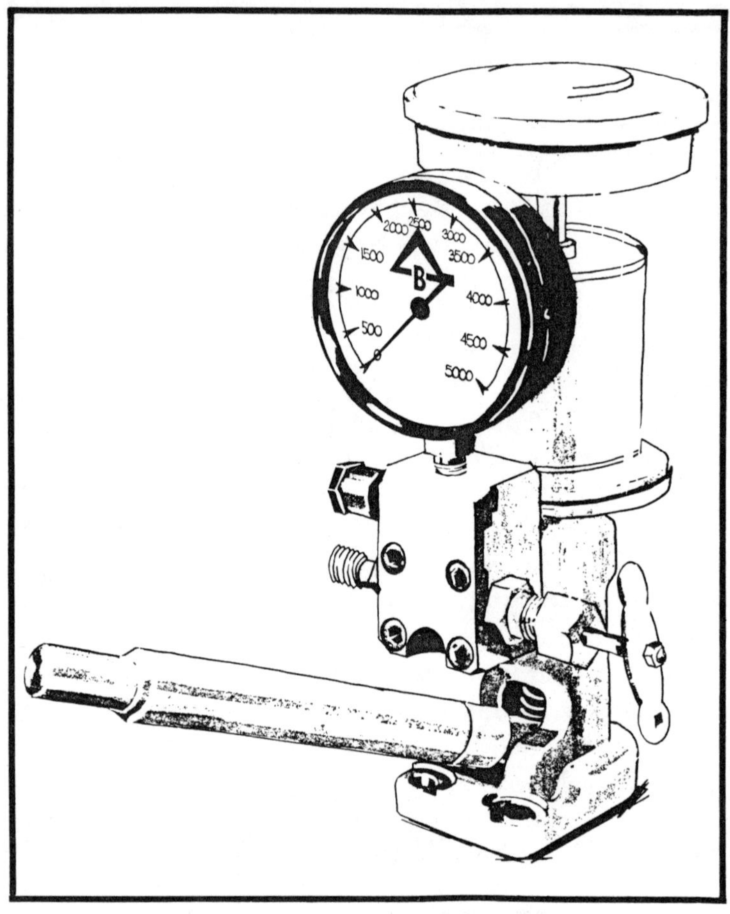

Fig. 5-38. A Bacharach series 65-0030 nozzle tester.

Run in the flange nuts finger-tight, then screw on the injector union nuts by hand, starting at the pump end. Tighten the union nuts and then the flange nuts to manufacturer's specifications. These specifications vary widely, so do not transfer torque values from one engine to another. If the union nuts leak, back off the threads and start over. Connect the return line.

Fuel Timer

The fuel timer is similar to the centrifugal advance mechanism on ignition distributors. On both gasoline and diesel engines, ignition should occur progressively earlier as engine speed increases. The rate of flame propagation is more or less constant and independent of piston speed. An engine that is timed correctly at idle will be late at full throttle. Peak chamber pressures will occur after the piston has dropped in the bore, wasting fuel and robbing power.

Figure 5-40 illustrates the Chrysler-Nissan timer. It is mounted on the inline pump between the pump drive shaft and the engine drive gear. As speeds increase, the timer advances the pump drive shaft a few degrees and initiates earlier injection and ignition.

Peugeot Deferred Injection

Deferred injection lengthens the injection period during idle. The same amount of fuel is delivered, but at a slower rate. Consequently, there is less fuel in the cylinders during ignition and combustion pressure builds slowly, quieting the idle. At higher-than-idle speeds, deferred injection would be inefficient and is cut out of the circuit.

Figure 5-41 illustrates the start of injection at idle. For the sake of clarity, only one of the four delivery valves is shown at the lower right of the pump. Fuel is subject to high pressure in chamber 7 and is pumped through ducts 5 and 6. The delivery valve is unseated by the pressure and fuel flows through the valve and to the injector. Valve 4 on the accumulator body is open and fuel passes into the accumulator where it compresses the spring above the plunger.

In the next drawing, Fig. 5-42, injection has begun. The accumulator spring is weaker than the injector spring, so some fuel does continue to bypass into the accumulator. In Fig. 5-43 injection

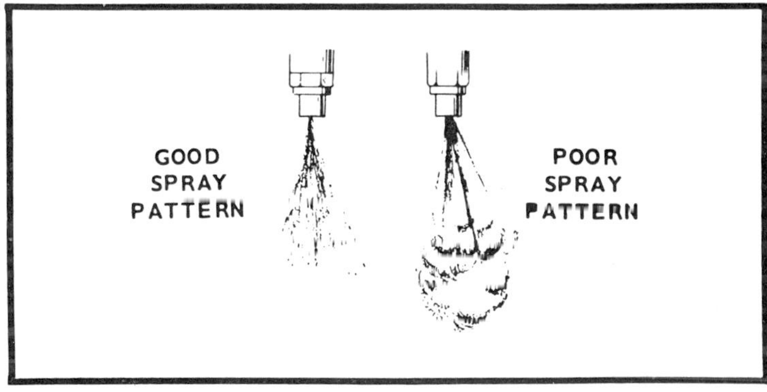

Fig. 5-39. Typical spray patterns for pintle nozzles. (Courtesy Onan.)

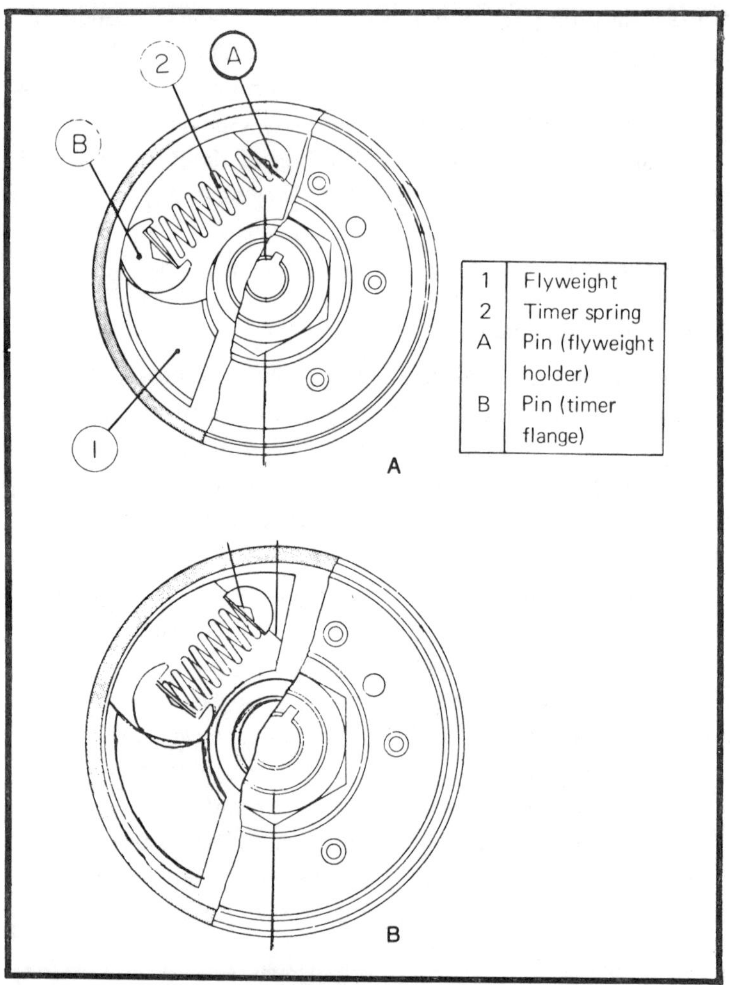

Fig. 5-40. A Chrysler-Nissan (Robert Bosch) fuel timer. At rest the spring keeps the flyweight retracted (view A); at high speed the weight moves, advancing the timer hub and pump drive shaft (view B).

has stopped; the groove (11) in the pump plunger has come in alignment with the spill port (10) and pressure in the chamber (7) falls rapidly. The delivery valve (9) seats, ending injection. The accumulator spring (1) pushes the accumulator plunger (2) down, discharging fuel back into the pump through the duct (11).

The accumulator can be thought of as a kind of hydraulic shock absorber, extending injection time by slowing the rate of pressure

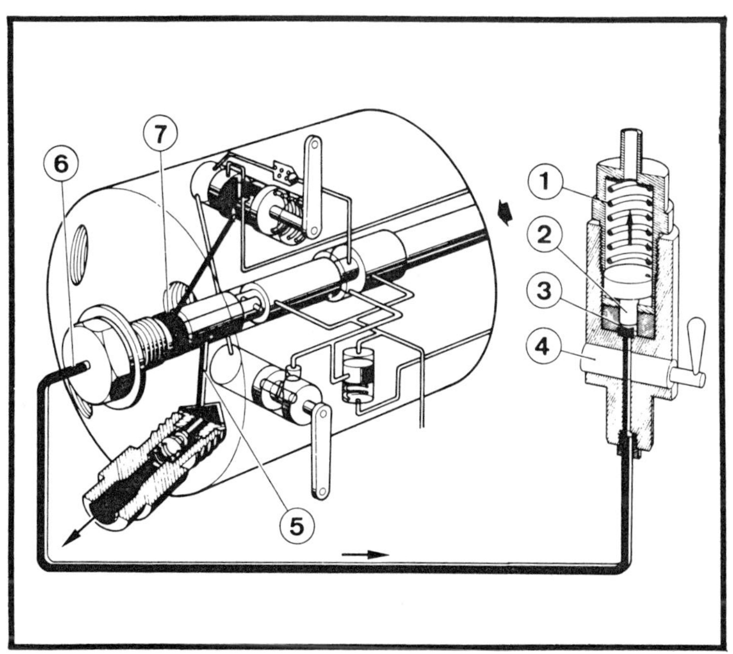

Fig. 5-41. Peugeot deferred injection at the beginning of injection.

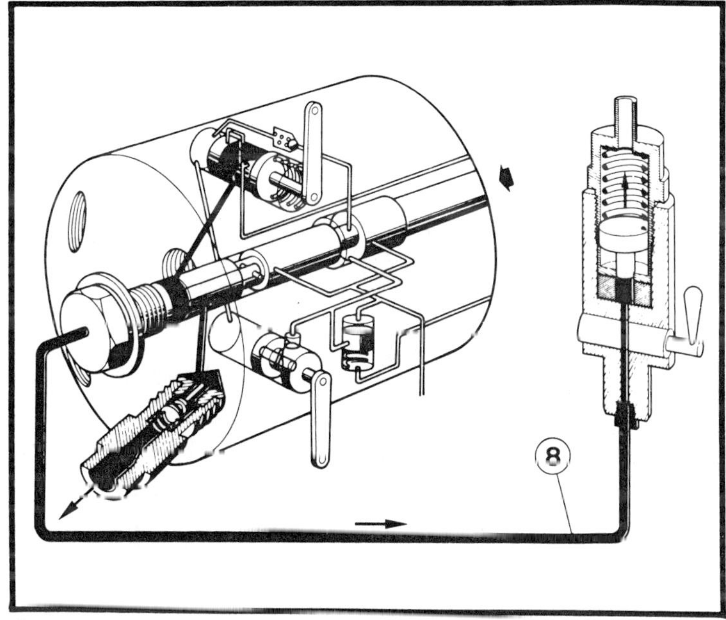

Fig. 5-42. At this point injection has commenced and the accumulator is storing fuel.

increase. The pump must operate at full capacity to deliver enough fuel.

GOVERNORS

Diesel engines have governors for three reasons:

1. To regulate idle speed.
2. To set an upper speed limit.
3. To hold the engine at any constant speed, between idle and maximum.

The governor moves the control rack in response to a signal generated by engine speed. The way in which this signal is generated, or how the governor "sees" engine rpm, varies. I will discuss the two most popular types, pneumatic and mechanical.

Pneumatic

Pneumatic, or "flap valve," governors are used on a number of automotive engines, including Mercedes-Benz. These governors sense engine speed as a function of air velocity through the intake header.

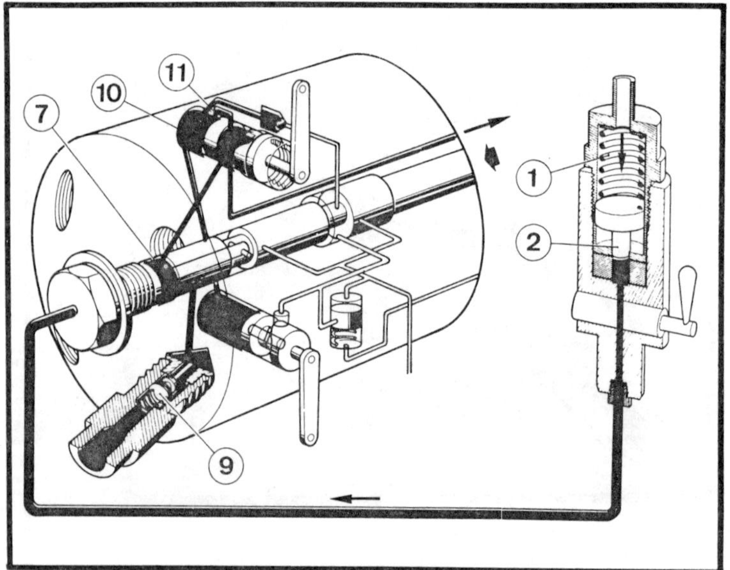

Fig. 5-43. The end of injection. The accumulator discharges back into the pump housing.

The header pipe is fitted with a spring-loaded flap valve, shown as part No. 4 in Fig. 5-44. The flap valve responds to air speed. When the engine is idling or operating at low speed, spring tension on the valve keeps it shut. Air going to the engine is diverted around the edge of the valve and passes through the Venturi channel (2).

A Venturi channel is a length of tubing that pinches in at some point, rather like an hour-glass, to a diameter between 1/3 and 3/4 that of the rest of the tubing. When air or liquid passes through a Venturi channel, it will speed up at the pinched-in section. This speeding up creates a region of lower pressure in the pinched-in throat. If you run a pipe from that throat to a vessel of some other air or liquid, and that air or liquid is at a higher pressure than you have in the throat, the contents of that vessel will be drawn through the pipe to the throat and into the tubing. This Venturi channel principle is used in carburetors and is even more familiar from its usage in flush toilets.

In our case, the positioning of the flap in the chamber as the air is drawn through creates the conditions for the Venturi-channel effect. The region of low pressure that results above and behind the shelter of the interfering flap is called a "vacuum," which term does not mean that there is *no* air, but that there is *less*.

The governor mechanism in Fig. 5-44 is connected to the header by two air lines. One brings filtered air to the right side of the diaphragm chamber. This air supply is at atmospheric pressure, less the pressure drop across the filter element and in the line. The second line conveys the vacuum signal to the left side of the chamber.

The chamber is divided by a spring-loaded diaphragm (8). The diaphragm is connected to the control rack, which is geared to the individual pump plungers. A second view of this mechanism is shown in Fig. 5-45. The variation of the rack movement changes the effective stroke of the plungers and the amount of fuel delivered has been discussed in the section on pumps.

So long as the flap valve remains stationary in the intake bore, vacuum in the left side of the chamber remains constant. The diaphragm takes a position determined by spring tension and the pressure differential between the two sides of the chamber.

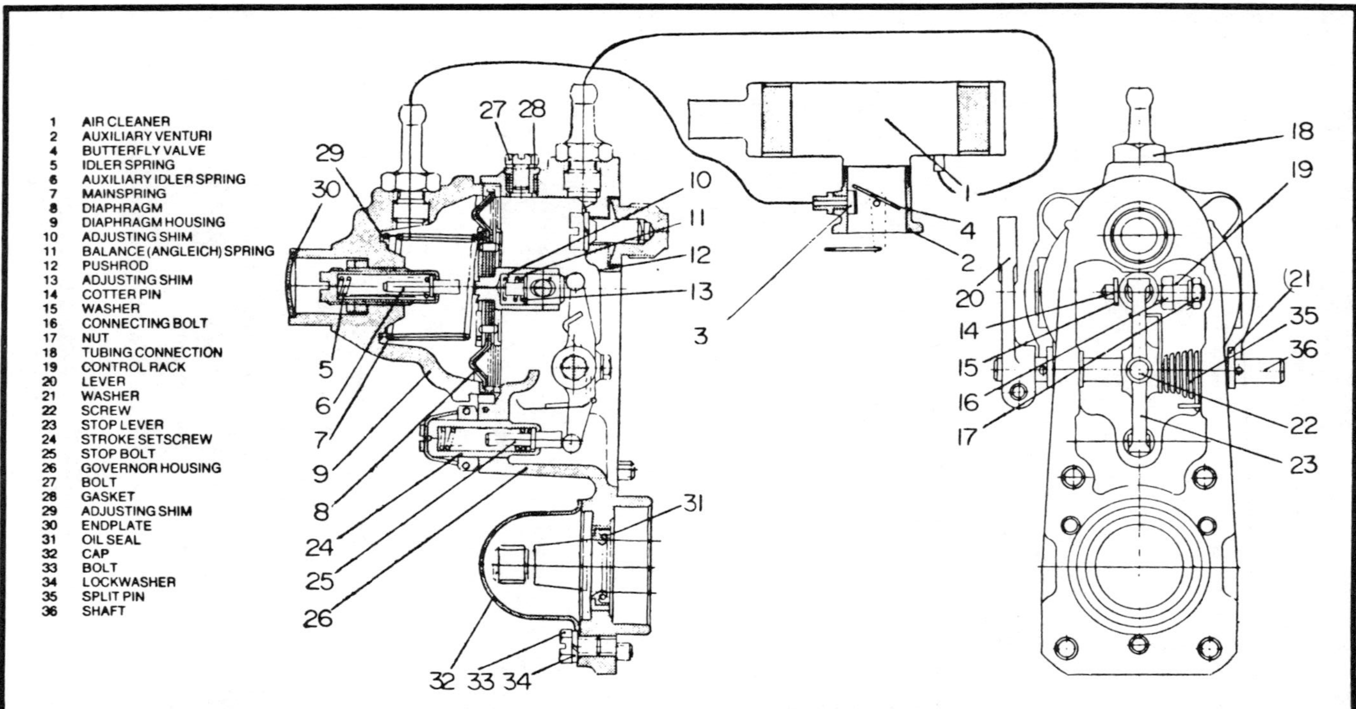

Fig. 5-44. The MZ pneumatic governor in cutaway. (Courtesy Chrysler-Nissan.)

As engine speed increases, the additional velocity of the air-charge causes the flap valve to open. Correspondingly less air is admitted to the Venturi channel and vacuum drops. The diaphragm, impelled by its spring, moves to the right. The attached control rack moves inward, reducing pump output until a new equilibrium is established.

Should engine rpm drop, as under sudden load, the flap valve moves toward the closed position. More Venturi vacuum is developed and the diaphragm moves to the left, pulling the rack out and increasing fuel delivery.

The stop lever (23) in Fig. 5-44 has two adjustments: the stop bolt (25) limits the maximum fuel delivery during cold cranking, and the stroke setscrew (24) limits fuel once the engine fires. The fuel enrichment provision is by means of a spring that temporarily overrides the stroke screw adjustment. This spring is depressed during cold starts.

The "angleich," which means "balance" in German, mechanism is shown in detail in Fig. 5-46. It is a corrective device that reduces fuel delivery at low rpm by moving the control rack farther out than would otherwise have been the case. At very low speeds, vacuum is diminished even though the flap valve is almost closed because the Venturi generates vacuum in response to the initial air velocity. Consequently, the diaphragm would move too far to the left and bring the control rack almost to the fuel shutoff position.

The angleich spring rides in a plunger on the end of the control rack and limits inward movement so long as the plunger contacts the stop lever. At higher engine speeds the diaphragm pulls the rack out, away from the stop lever. The spring is still in the circuit, but is not compressed by diaphragm movements to the right. The only resistance against the rack is the reluctance of the plungers to turn, and this is inconsequential, amounting to a few grams.

The MV governor is described here. It differs little from other makes. All employ a velocity sensitive flap valve, a Venturi to generate vacuum, and a diaphragm that moves in response to the vacuum signal. However, most other makes are simpler than the MV. The Simms does not use an angleich spring and the CAV ports the high pressure side of the chamber directly rather than through a second line to the intake air filter.

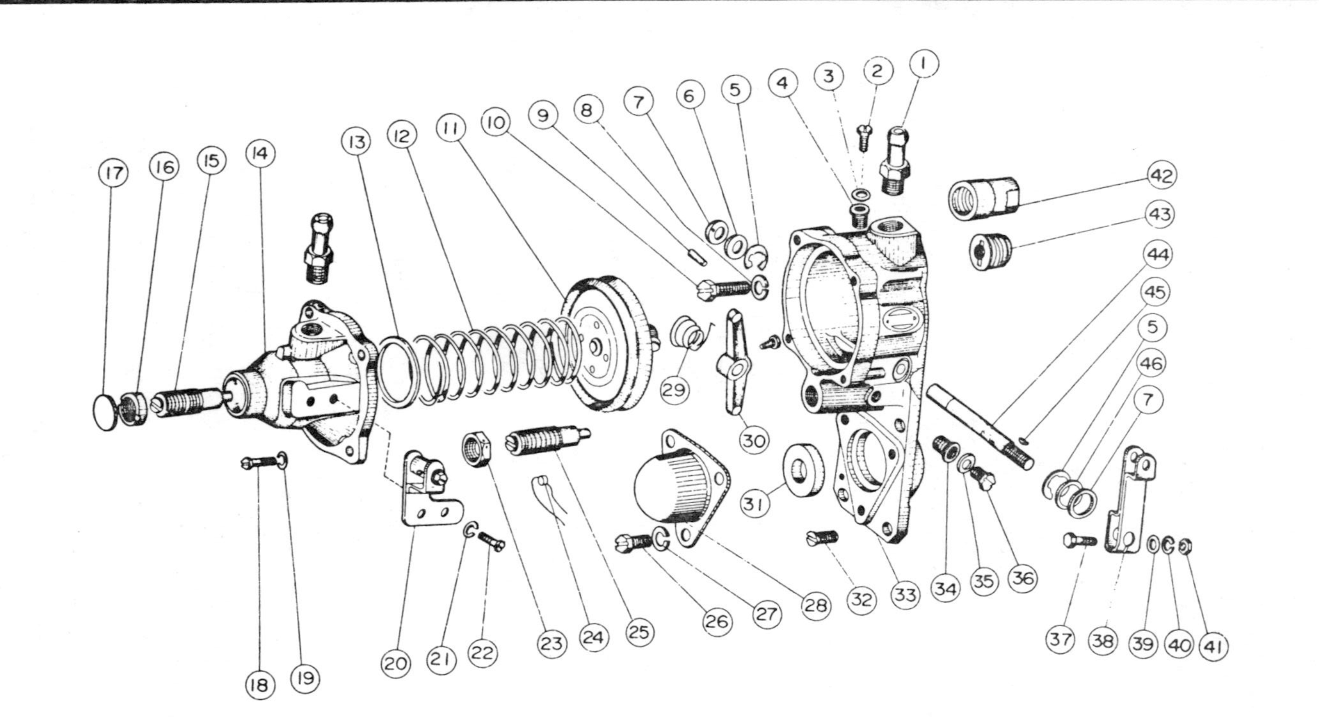
1
2
3
4
5
6
7
8
9
10
11
12
13
14
15
16
17
18
19
20
21
22
23
24
25
26
27
28
29
30
31
32
33
34
35
36
37
38
39
40
41
42
43
44
45
5
46
7

1	Pipe joint bolt	24	Lead seal
2	Bolt	25	Spring guide assembly
3	Gasket	26	Screw
4	Nipple	27	Lock washer
5	Snap ring	28	Cap
6	Adjusting shim	29	Return spring
7	Washer	30	Stop lever
8	Lock washer	31	Oil seal
9	Guide pin	32	Screw
10	Screw	33	Governor housing
11	Diaphragm	34	Nipple
12	Governor spring	35	Gasket
13	Adjusting shim	36	Screw
14	Diaphragm housing	37	Bolt
15	Spring guide assembly	38	Control lever
16	Round nut	39	Washer
17	End plate	40	Lock washer
18	Screw	41	Nut
19	Lock washer	42	Cap
20	Clamp assembly	43	Screw plug
21	Spring washer	44	Shaft
22	Screw	45	Key
23	Lock nut	46	O-ring

Fig. 5-45. The parts relationship in the governor proper. (Courtesy Chrysler- Nissan.)

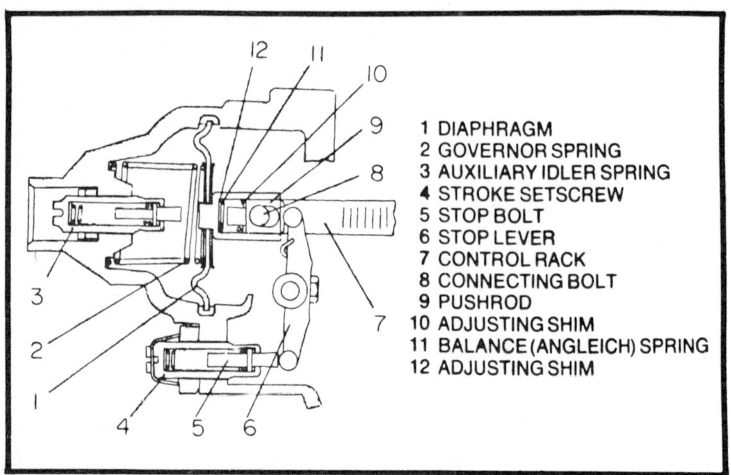

Fig. 5-46. The balance mechanism. (Courtesy Chrysler-Nissan.)

Service

A pneumatic governor requires no routine maintenance other than checking that the line fittings are tight. After long service the flap valve shaft may wear enough to throw the governor out of calibration at low speeds. Replacement shafts and bushings are available. The leather diaphragm may grow stiff in service and give erratic response. It can usually be salvaged by application of neats-foot oil, which may be sold at engine outlets as "diaphragm oil." New gaskets which have been coated with liquid silicone or some other trustworthy sealer must be used whenever the governor unit is disassembled. By the same token, the line fitting threads should be coated with Loctite. A vacuum leak will be interpreted by the governor as in sufficient fuel and the engine will continue to gather speed until it disintegrates.

Mechanical Governors

Mechanical, or centrifugal, governors are developments of the flyball governors used on steam engines. The governor is bolted behind the pump and driven by the pump shaft at half engine speed, on four-cycle engines.

Figure 5-47 is a cutaway of the Simms governor. While the Simms is as complicated as it looks, it is a masterpiece of simplicity compared to some other makes. The housing (A) encloses the mechanism and serves as an oil sump. This example has its own oil supply but others share the cambox lubricant or may be connected to the engine oiling circuit.

The pump shaft (B) rotates the weights (C). The weights pivot outward and move the yoke and socket (G) to the left, causing the crank lever (D) to move against spring tension. Crank lever motion is conveyed to the control rack (J) shown at the upper right. Spring E is known as the idling spring. Because it is mounted on articulated links, spring E comes into play only during idle when centrifugal forces are weak. Spring F controls response at high speed. The throttle lever is secured by nut H.

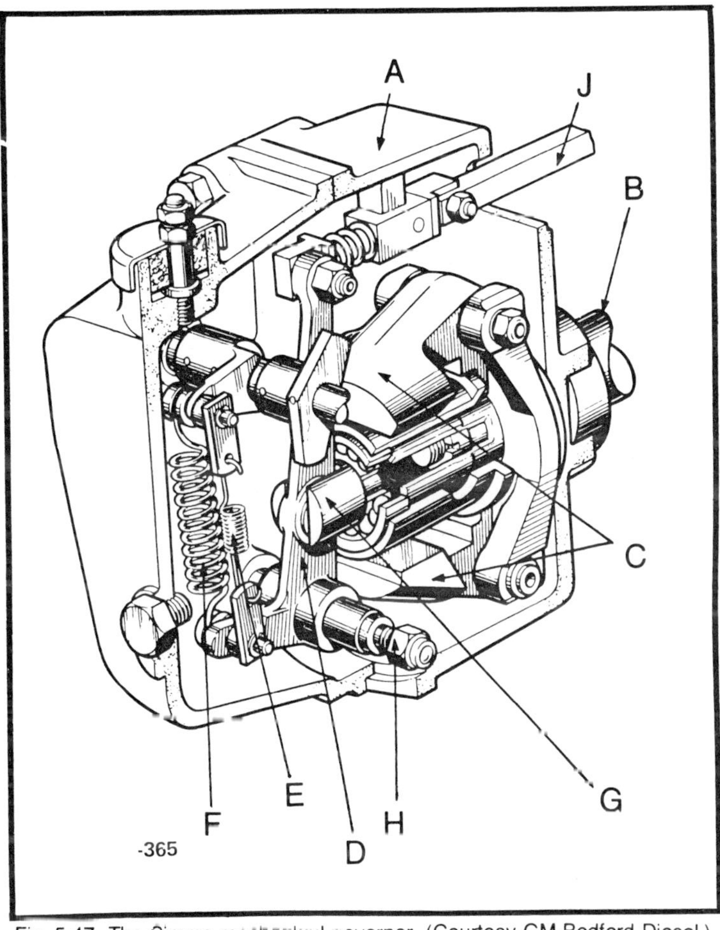

Fig. 5-47. The Simms mechanical governor. (Courtesy GM-Bedford Diesel.)

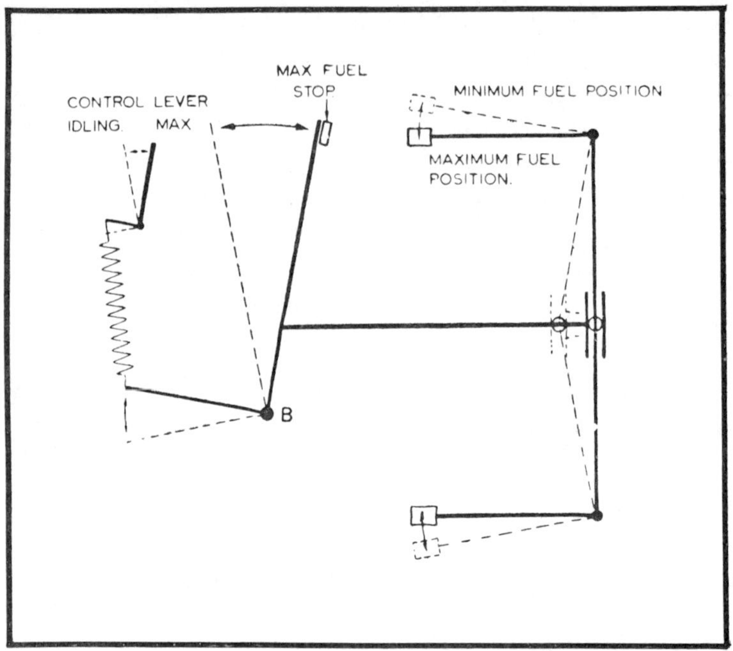

Fig. 5-48. The chain of motion in the Simms governor. (Courtesy GM-Bedford Diesel.)

Operation is shown diagrammatically in Fig. 5-48. During start-up the weights are relaxed and the control rack is moved to the maximum fuel position. The heavy line represents the articulation of the parts. Once the engine catches, the weights move outward, reducing fuel delivery. The point at which centrifugal force and spring tension stabilize is determined by the spring loading, a function of the throttle lever setting.

While all centrifugal governors operate on the same broad principle, there are differences in details between makes. The CAV governor is shown in simplified form in Figs. 5-49 and 5-50. These sketches point out some peculiarities of the instrument. One important feature is that spring loading is constant and independent of throttle position. Another is that the weights move vertically on posts, rather than arc on heavily loaded pivot bearings.

The throttle lever (8) acting through arm 11, pivots the control lever (1) on axle 12. A second axle (2) allows independent movement of the control lever in response to the yoke (4) and the flyweights (6). The flyweights react against the springs (13).

Figure 5-49 shows the mechanism set at midspeed. Should the engine fall off under load, centrifugal force diminishes, and the weights move together. This movement sends the yoke to the left, pivoting the upper control arm end and the fuel rack rightward for more fuel. Under undesired acceleration the chain of action is reversed: the flyweights move out against their springs, the yoke pulls to the right, and the control arm pivots to the left, toward the stop position. The axle (12) plays no part in these automatic responses, and indeed is locked in place by a pin-and-hole arrangement on the lever quadrant. In an automobile, the throttle would be positioned by disadvantageous leverages or, in a first-class conversion, by the use of a hydraulic throttle.

The next drawing, Fig. 5-50, illustrates how the three-quarter speed position shifts the control linkages to the right on the axle (12). The weights have not been affected, and continue to act on the yoke through the articulated arms. The rabbit-eared springs (9) are a failsafe. The limits of throttle lever movement are defined by the stop screw and the maximum fuel delivery screw. These screws are accessible from outside the governor, and can be maladjusted. If this were the case, extreme movements of the throttle lever would engage either spring, rather than stress the linkages.

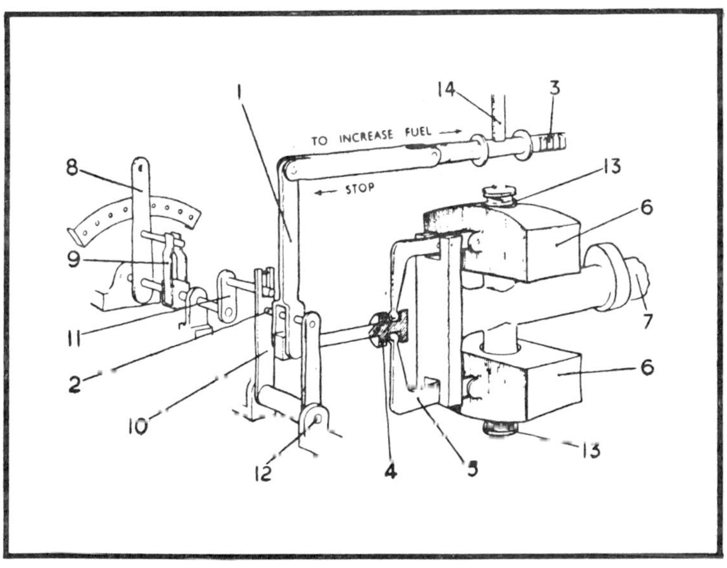

Fig. 5-49. The Simms governor at half-speed. (Courtesy GM-Bedford Diesel.)

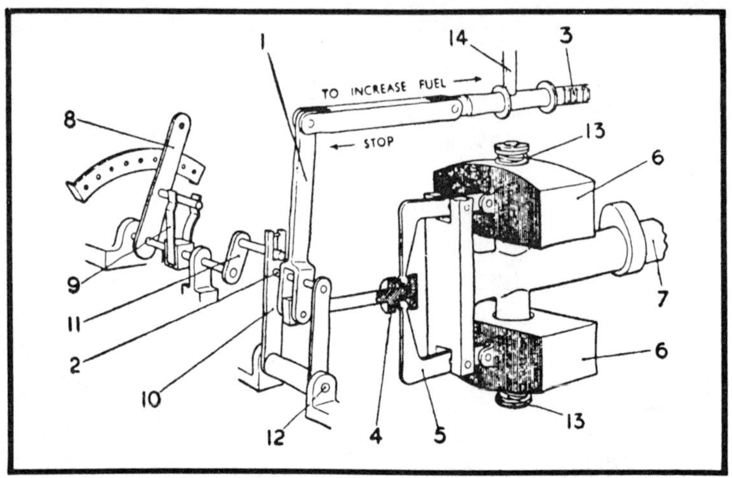

Fig. 5-50. The Simms governor at full-speed. (Courtesy GM-Bedford Diesel.)

Repairs

Governor malfunctions such as hunting, sticking, slow or erratic response can usually be traced to sticky or worn pivot bearings. These governors should hold engine speed within 2% of the throttle setting under no- and light-load conditions.

The first line of defense is to drain the oil and remove the cover. Clean the parts with trichloroethylene or with one of the aerosol gum-cutting products such as Berkerly 2-2 Gum Cutter. Work the mechanism by hand to introduce the spray into the bearings. If this is ineffective or if the bearings have perceptible play, replace the affected bushings and pins. Complete disassembly requires special tools and an inventory of shims which make this work impractical for the average owner.

SUMMARY

In conclusion the fuel system is the most complex and technologically demanding of all diesel systems. Repair operations should be carried on with discretion and under conditions of extreme cleanliness. There is a limit to field repairs which is imposed by special test apparatus, parts inventory, and experience. IHC and other manufacturers underline this limit by omitting fuel system data from their engine manuals. Ordinary line and bench mechanics do not have access to this information. Those who do presumably have the expertise to make correct use of it.

Overhaul And Rebuild

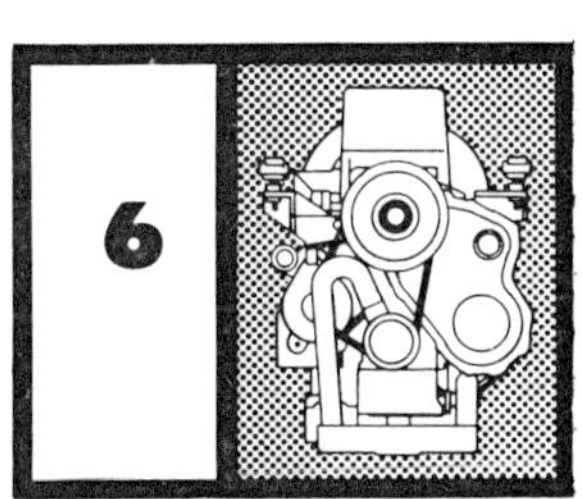

Diesel engines are mechanically similar to gasoline engines, but built on a grander scale. The same general overhaul and rebuild techniques apply to both. However, because of the higher bearing pressures in diesel engines and because of the presumption of durability, diesel engines do not tolerate many mistakes. Work in a diesel is slower, more demanding, and more rewarding.

PRELIMINARIES

You will need a chain or cable hoist, preferably mounted on a wheeled A-frame. The hoist adapter varies with make and model, but it should incorporate a spreader bar to keep the lifting forces vertical (Fig. 6-1). Most engines have lifting tabs fore and aft. You can easily fabricate a pair of tabs from 1/4 inch or heavier steel plate, bent at right angles and drilled for the headbolt and sling hook. In some cases, engine balance is best served if the intake header or exhaust manifold is used as one attachment point.

An engine stand is a great convenience, but the lack of one should not impede the work. Note that the stand illustrated in Fig. 6-2 follows diesel practice and mounts to the side of the engine, allowing almost complete accessibility.

The external castings should be clean before disassembly or it will be almost impossible to keep dirt out of the engine once it is

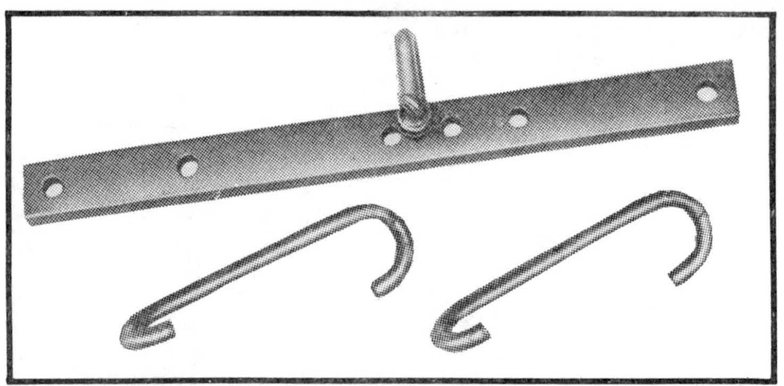

Fig. 6-1. Spreader bar and lift hooks can be easily fabricated.

disassembled. Remove the starter motor and alternator, together with the air-filter element. Seal the air intake with duct tape and cap off the fuel lines. Drain the oil and snug down the plug, as if the engine were immediately going back into service. I insist on this because newly rebuilt engines have been lost when a finger tight drain shook loose.

No diesel stays clean very long, but vehicle engines are the dirtiest of the lot. If this is your first experience with diesels, you'll be unpleasantly surprised by the way diesel dirt propagates. It's like printer's ink—a smudge on your hand will leave prints on everything that you touch.

The best way to clean an engine is with live steam and detergent. The next best way is to haul the thing to a car wash that is equipped with a boiler and high pressure pumps. Another method is to douse the engine with trichloroethylene. Unfortunately this industrial solvent is difficult to purchase in less than 55 gallon lots and may be a health hazard. One lesser side effect of inhaling trichloroethylene is known as "degreaser's flush," and it is embarrassing to say the least that after prolonged exposure to trichloroethylene a single shot of alcohol turns your face bright red.

The old standby is Gunk, "the original self-emulsifying degreaser." When mixed with kerosene, Gunk turns oil and grease into a low-grade soap that can be washed off with a garden hose. Gunk will soften paint but it can do little against baked-on grease such as you will sometimes find around the exhaust manifold and on Deutz cooling fins. For that, the best bet is oven-cleaner.

ROCKER ARMS

Remove the rocker arm cover and turn the engine over a few times to check valve lash. If value lash is excessive, you can be sure that the engine was skimpily maintained and you can only hope that the first owner kept fresh oil in it.

Remove the rocker-arm assembly, working from the center shaft bolts out. A few turns at a time on each bolt is enough until spring tension is relieved. Note which, if any, bolts are drilled or grooved for oil circulation. Lift the rocker assembly clear. If there is any possibility of confusion, as there might be with early Mercedes engines that use two rocker shafts, mark the assembly that is toward the front of the engine.

Lift out the pushrods, placing them in a numbered rack so they can be replaced in their original sequence. Inspect the tips for obvious, thumb-nail hanging wear. Check pushrod straightness by

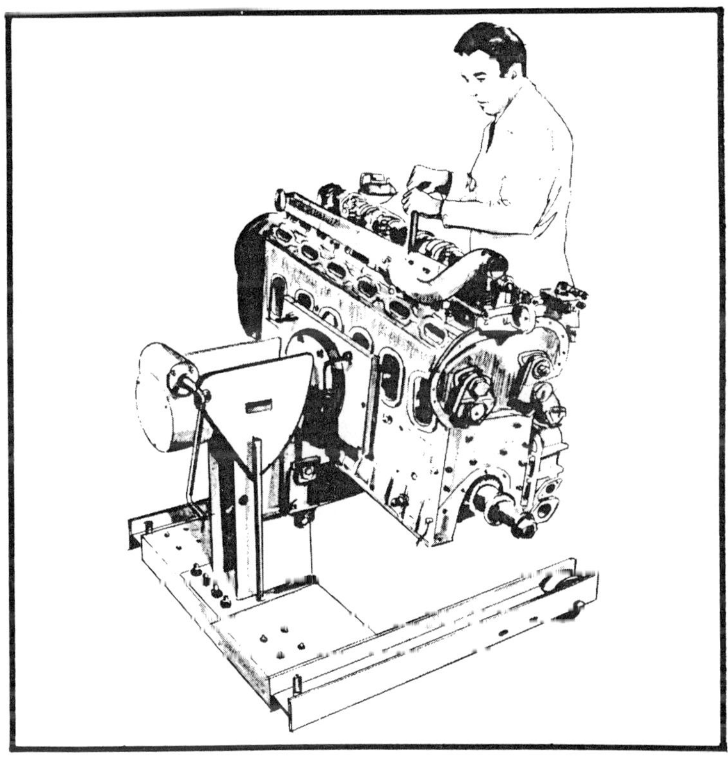

Fig. 6-2. A Bacharach (code 82-7012) engine stand; the ultimate in convenience.

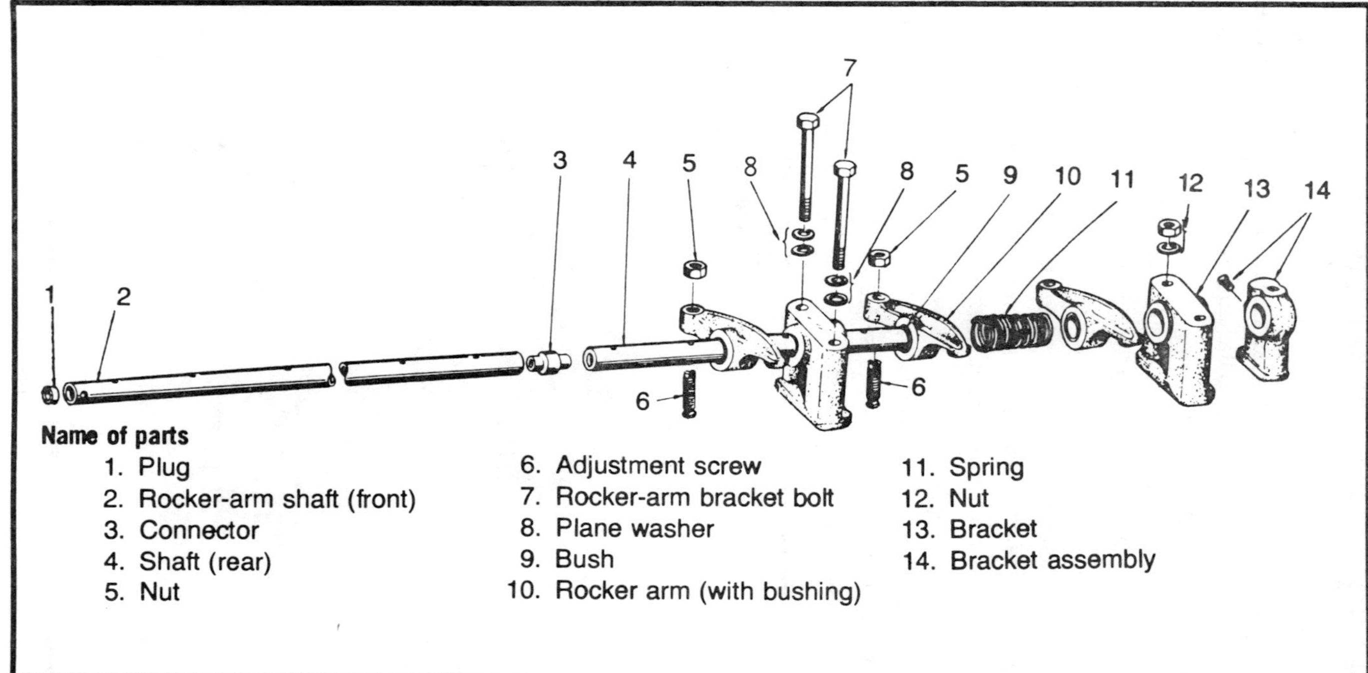

Fig. 6-3. Isuzu rocker-arm shaft and related hardware.

rolling the rods on plate glass or on a machined work table. Replace any that wobble when they roll.

Figure 6-3 illustrates an Isuzu rocker-arm assembly. It is typical of most in that the rocker shaft serves as an oil gallery with cross-drilled holes for oil distribution to the individual rockers. It is typical of all in that there is no interchangeability of parts for a used assembly. Each rocker, spring, pedestal, and shim must go back together exactly as originally assembled. The parts should be laid out on the bench in order.

Pull out the cotter pin at either end of the rocker shaft, together with the flat and wave washer. Withdraw the adjacent rocker arm and inspect it very carefully. Look for:

1. Wear on the adjustment screw. The spherical end should hold its original profile, without tell-tale "splash" marks. Loosen the locknut and try the threads. The screw should turn easily and smoothly. Self-locking screws are supposed to turn hard, but they should not bind solid.
2. Wear on the rocker-arm face, where the arm rests against the valve stem. Minor wear is usually ignored, although it is worthwhile to have the rocker dressed to its original profile. Some automotive shops have the equipment to do this, and the cost is nominal. The reward is a quieter engine that holds its lash adjustment longer.
3. Cracks radiating out from the bushing. This sort of failure is rare, but unfortunately not unknown. Some builders Magniflux rocker arms for insurance.
4. Bushing wear. Clearance between the bushing and the unworn part of the shaft should be no more than 0.002 in. New bushings are pressed in and finish-reamed. Take care to align the oil hole with the hole in the rocker. On some designs the bushing is split; the gap is indexed with the rocker port and acts as a mini-gallery to distribute oil over the whole length of the bearing.

So much for one rocker arm.

Withdraw the first pedestal. If it is stubborn, do not force the issue. Instead, wrap the pedestal with a shop towel and pour hot (150°F) oil over it. Allow a few minutes for the pedestal to heat and grow, and it will slide right off. Check for radial cracks around the bolt holes and, if an oil port is present, clear it with compressed air.

Continue to dismantle, cleaning and inspecting each part as it comes off, and laying them out in their assembled order. The shaft will be the last to be inspected. Unless a rocker bushing has died from oil starvation, wear will be confined to the underside of the shaft. It is difficult to generalize about how much wear is permissible, but if the shaft is 0.002 in. out-of-round, you can always use it for a pry bar. Detect bends by rolling the shaft on a flat surface, just as you did with the pushrods. Some wobble is allowed, but be leery of a shaft that is more than 0.006 in. out.

If replacement end plugs are available, it is a good idea to uncap the shaft for a thorough cleaning. Otherwise, soak it in carburetor cleaner and rinse with solvent.

Once you are satisfied that the assembly can give you the service expected of diesel parts, coat the rubbing surfaces with clean lube oil and assemble.

Now, while the head is still attached we have not yet gotten down to the moil and toil of repair, it is a good time to start thinking about the oiling system. Checking out this system is one of the most important things that you will be doing, and is justification in itself for opening an engine with much time on the clock. Figure 6-4 outlines the Peugeot XDP 88-90 circuit. In principle, it is similiar to others in that oil is picked up in the sump, passes through a series filter and enters the main oil gallery running longitudinally along the side of the engine. The gallery is the central distribution point which supplies oil to the crankshaft via the main bearing webs, to the valve lifters, and to the rocker arms. In this case, rocker arm feed is by means of an external line. Oiling systems differ in detail between makes and models. It is important to be aware of their individual idiosyncrasies, so that each element of the system can be inspected. Figure 6-5 shows a Chrysler-Mitsubishi circuit. It is not surprising that this engine is considered the class swap. Detroit Diesel piston pins are covered with disc-shaped retainers to prevent oil from bleeding on the cylinder walls. If these retainers leak, the engine will consume oil. Turbocharged IHC engines cool the undersides of the pistons with jets of oil directed from the sump.

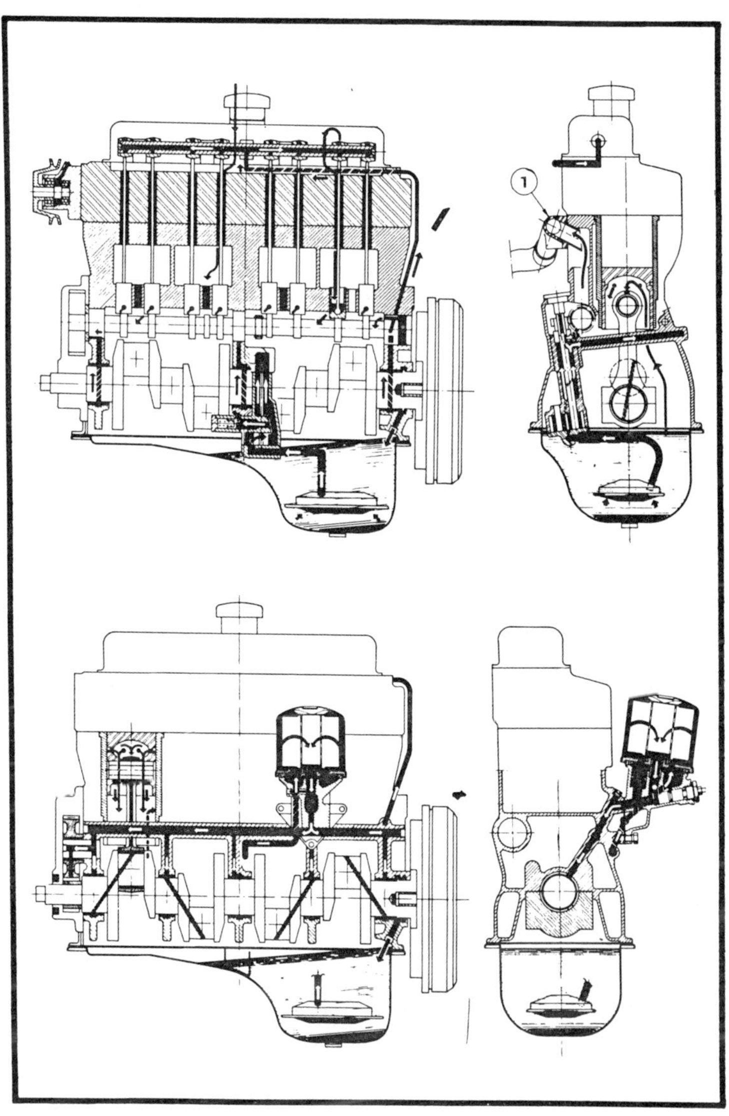

Fig. 6-4. Peugeot oiling circuit. As is often found in diesel engines, the valve mechanism feeds from an external line.

CYLINDER HEAD

Prepare to lift the cylinder head. Remove these items:

- Intake header
- Exhaust manifold

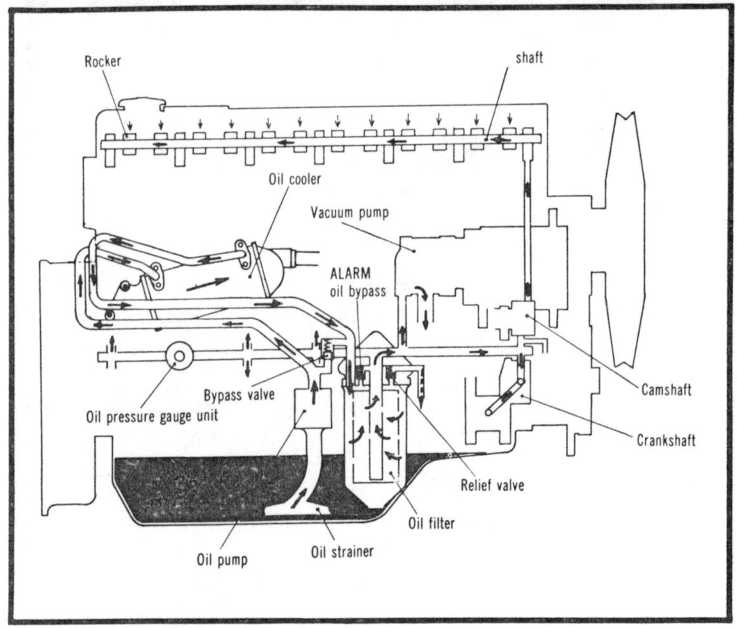

Fig. 6-5. Late—after 6DR5-7405—Chrysler-Mitsubishi engines use this setup with oil cooler, filter bypass alarm, and vacuum pump in the circuit.

- Injectors
- Miscellaneous gear—vacuum pump, alternator, governor signal and oil line, etc.

Warning: Intake, exhaust, and cylinder head gaskets are made of asbestos. The dust from these gaskets can be lethal if ingested or inhaled. Scrape the gasket surfaces; do not use a wire brush. For greater safety, you may even want to wear a mask. They are inexpensive and readily available.

It is good practice to have the injectors tested whenever they are removed. Inspect the intake header and exhaust manifold for cracks and dress the flanges with a file. Move the file from left to right over the flange until the metal takes a uniform luster, meaning that it will be dead flat.

Using a breaker bar, remove the head bolts. Because diesel cylinder heads are "busy" with bolts, valve assemblies, and injector shells, all the head bolts may not be interchangeable; for example, three headbolt lengths are used for the M-D 300. Detroit Diesel engines normally have 6-point head bolts, except those near the camshaft which are 12-point for the additional clearance. Nor are the bolts always the same length. As you remove the bolts, make note of any that have been coated with water-resistant sealant. Clean the bolts in solvent, wire-brushing the threads by hand as necessary.

Get help to lift the head. Diesel castings are as heavy as they look, and hernias and back injuries are expensive and painful. If you have time, nail together a holding fixture along the general lines of the one shown in Fig. 6-6. Lacking this, at least support the head on wood blocks. Gasket surfaces are vulnerable and some heads have fragile swirl-chamber tips that extend beyond the parting line.

Scrape the carbon from the piston crowns and the combustion chambers. Be very careful not to scratch the gasket surfaces. Inspect the head and block for obvious cracks, particularly between the valve seats and cylinder bores. If you find this kind of damage,

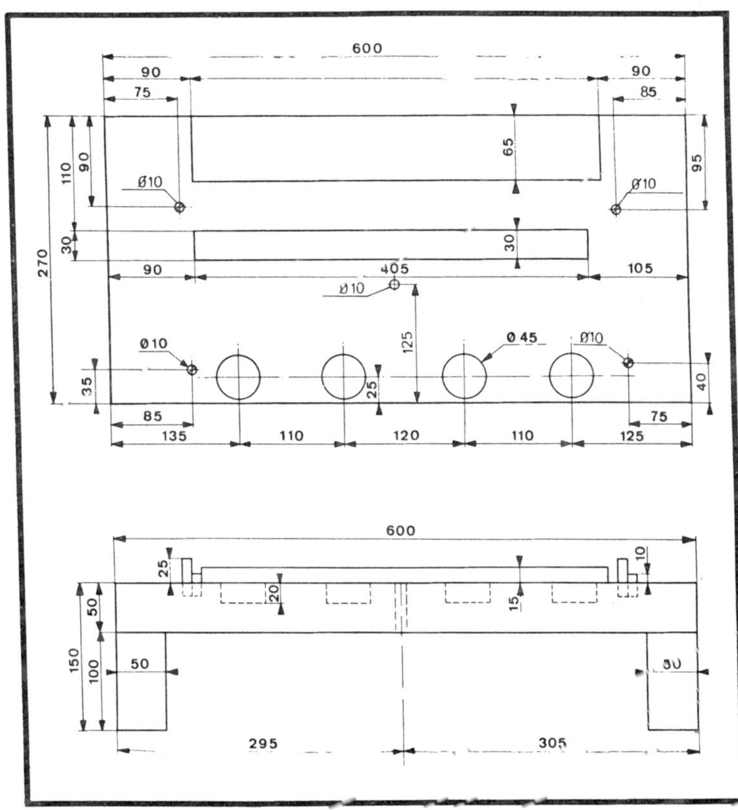

Fig. 6-6. Peugeot head-holding fixture.

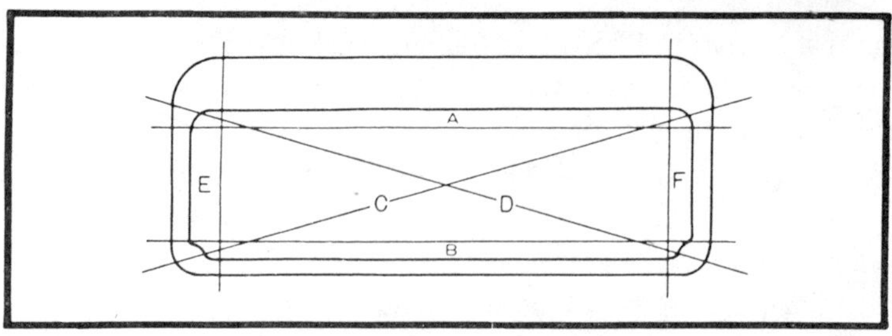

Fig. 6-7. Check cylinder head trueness with a machinist's straightedge, laid diagonally (C and D) and across (E and F) the gasket surface. (Courtesy Isuzu Motors, Ltd.)

you should consider another engine. For even if the crack is welded or the casting replaced, you can figure that the engine was abused and that if the crack extends into the water jacket there will also be lower end problems.

Lay a machinist's rule lengthwise across the head gasket surface to determine the amount of warpage (Fig. 6-7). Some distortion is inevitable, but 0.005 in. on the long axis and 0.003 in. athwart the head is plenty. Much more than that and the head must be planed. Unfortunately, there is a limit to how much metal can be removed, for heads can tolerate only light skimming on the order of 0.020 in. without failing the minimum thickness specification (Fig. 6-8). The amount of metal removed should be stamped on the water jacket. Since machinists sometimes overlook this nicety, it is wise to mike the head of a used engine before any machine work is contemplated.

FREEZE PLUGS

Careful mechanics like to replace the freeze, or core, plugs while the head is off and the plugs are accessible. The only complica-

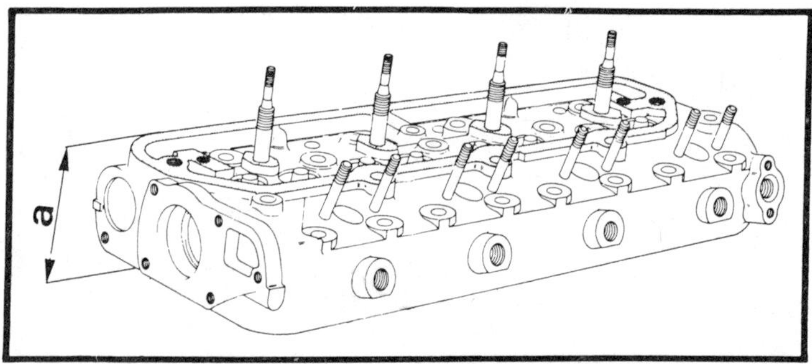

Fig. 6-8. Dimension A must meet or exceed factory thickness specs.

tion here is that most diesels are built on the International Standard (IS) and therefore require *metric* freeze plugs which small dealerships may not stock. Replacement plugs for American engines are available at auto parts houses

OEM plugs come in two styles—cup and expansion. Cup plugs must be installed with the proper tool (Fig. 6-9) one that does not contact the plug flange. A squared-off driver will cause plug distortion, leakage, and possibly blowout. A driver for expansion plugs must contact the outer edge of the plug and not the convex crown. The plug bores should be clean and coated with liquid silicone or some other waterproof sealant.

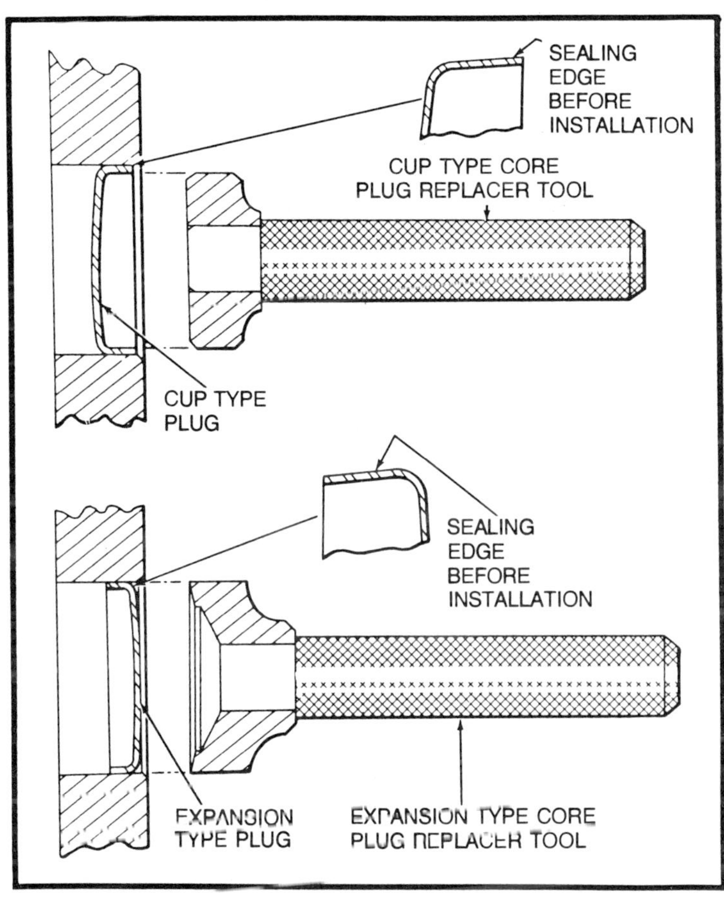

Fig. 6-9. These tools are rarely seen outside of Ford dealerships, but take the hassle out of freeze-plug installation.

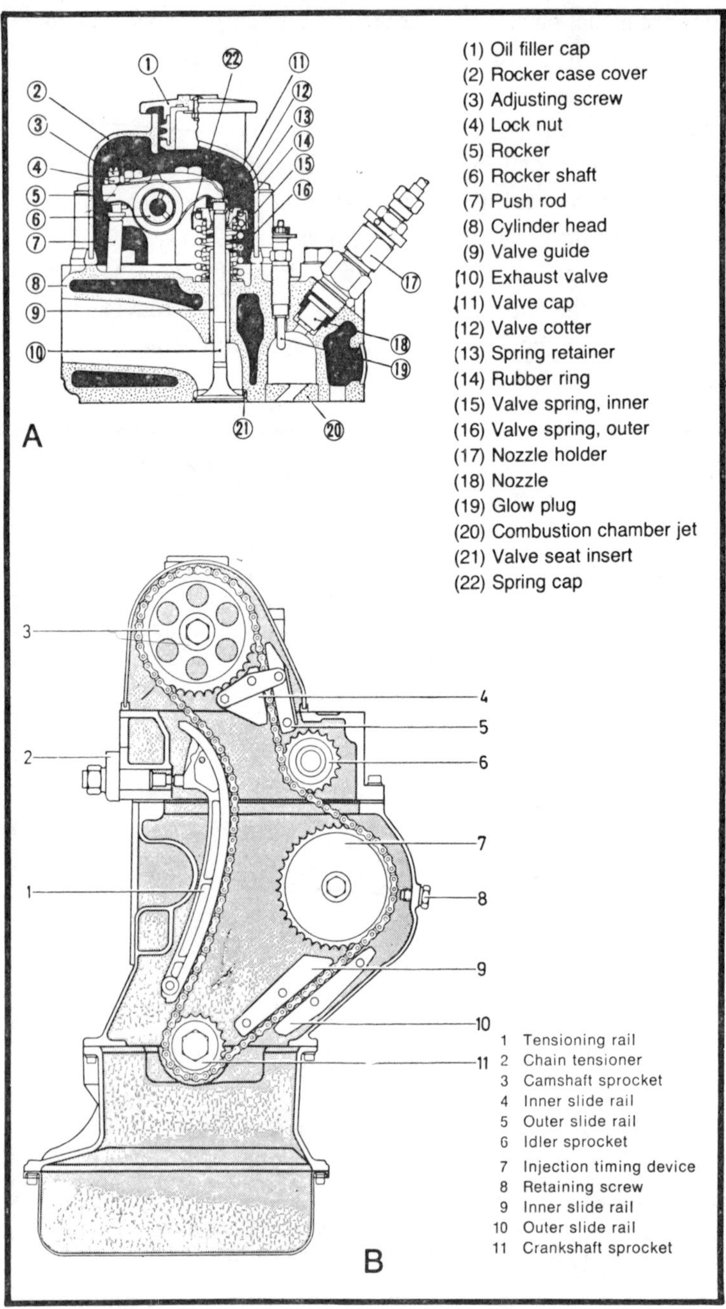

Fig. 6-10. Valve train varieties: the Mitsubishi overhead valve configuration (view A) and the Mercedes-Benz overhead cam setup (view B).

CORE TEST

The cylinder head core should be pressure-tested if you have installed new plugs or if you have reason to suspect coolant leaks. While pressure testing is almost unheard of in ordinary automotive work, it is a standard operating procedure among diesel mechanics. Use dummy injectors to seal the sleeves and cover water inlet and outlet passages with gasketed plates. Pressurize the core to about 40 psi. Immerse the head in hot water, at least as hot as the thermostat rating. After 20 minutes or so, check for bubbles.

Iron heads can be arc welded with a high-nickel content rod. Consult your supplier. Cast aluminum heads are tricky, but have been successfully patched with a Heliarc machine and a No. 2101-E Eutectrode.

VALVE MECHANISMS

The valves are operated from a camshaft low in the block on most engines. Movement is transmitted from the cam, via pushrods and rocker arms (Fig. 6-10A). Post-180 Mercedes engines employ a chain-driven camshaft mounted over the valves and working through "fingers" (Fig. 6-10B).

Valve Work

Machine grinding is the only way to accurately resurface the valve faces and seats. Hand-lapping is a waste of time for the most part, and can be positively detrimental if carried to extremes.

Valve-seat stones center on the valve guides. If the guides are worn, the stone wanders and wobbles, making a gas-tight seal impossible. Guide wear-limit is usually 0.005 in. on the intakes and 0.001 in. on the more critical exhausts. Most mechanics estimate wear by the amount of metal lost on the valve stems and by valve wobble, taken at the point of maximum valve extension. A better method is mike the valve stem and gauge the valve guide.

Worn valve guides can be corrected in any of three ways:

1. Light wear responds to knurling, a process that adds miles to guide life because it leaves oil reservoirs in its wake.
2. Heavily worn guides can be reamed oversize, but this entails new, special-order valves with matching stems

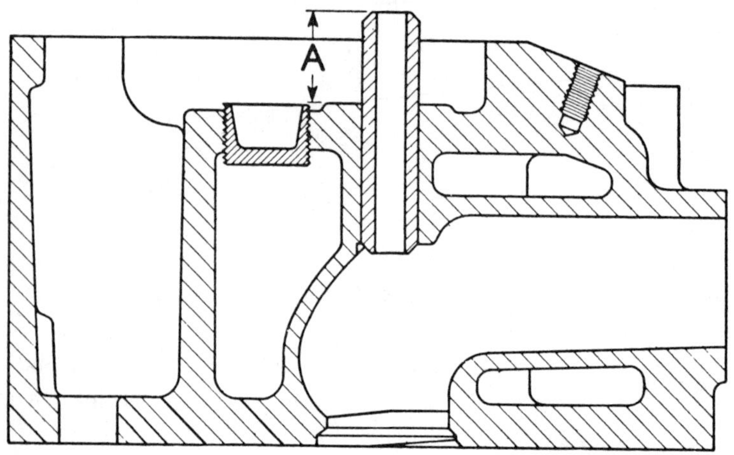

Fig. 6-11. Waukesha guides are not shouldered; height must be established by measurement or by reference to the published dimension.

These valves are expensive and slow to be delivered unless the dealer is particularly well inventoried.

3. Finally, the valve guide can be replaced.

Most mechanics avoid using oversized valves, but commercial engine rebuilders sometimes do. It is important that you check valve stem diameter against factory specs when replacing the guides.

To replace the guides, drive the old ones out and press in the new ones from above. Most engines have the guide depth fixed by a counterbore. Others do not have this feature and the depth of the original is the standard (Fig. 6-11). Once installed, the guide must be reamed to standard.

Valve Seat Inserts

Valve seats must be replaced if they are cracked, loose in their bores, or worn so much that regrinding would sink the valves. On many engines the seat overlaps the port, leaving a ledge on the underside that can be used to pry the seat out. Where the outer diameter of the seat does not extend beyond the port, the seat must be cut out. The cutter is slightly smaller than seat outside diameter, and leaves a narrow "rind" that can be easily broken (Fig. 6-12).

Once the seat is out, take a hard look at the counterbore. The counterbore must be true set at 90 degrees to the chamber without one of the tell-tale marks that mean insert wobble, and dead flat on the bottom. An inference fit is all that holds the seat insert in its counterbore. Most manufacturers supply oversized seats and the appropriate cutters. Without these parts, worn counterbores can be sometimes refurbished with oversized seats from another engine.

Factory seat drivers are available, although some mechanics use a discarded valve. Heat the head with a propane torch, moving the flame in circles over the counterbore. When the head begins to smoke, it is hot enough. Meanwhile chill the new insert in dry ice and alcohol or with Freon. Holding the insert with pliers, position it over the counterbore and drive it solidly home.

The valve-seat angle, the angle formed by that part of the seat that mates with the valve, is usually 30 degrees on the intake side and 45 degrees on the exhaust. This is because most pumping losses occur in the intake track; only 30% of them involve the exhaust system. A 30 degree seat flows 20% better than a 45 degree seat of the same diameter. However, the 45 degree seat generates a 20% higher seating pressure. The intake seats are angled for best flow, while the exhaust seats are angled for greatest pressure. The exhaust valves need all the help they can get to remain gas-tight, even if that help is at the expense of some flow restriction. The intakes run cooler and cleaner, and restrictions are of more moment.

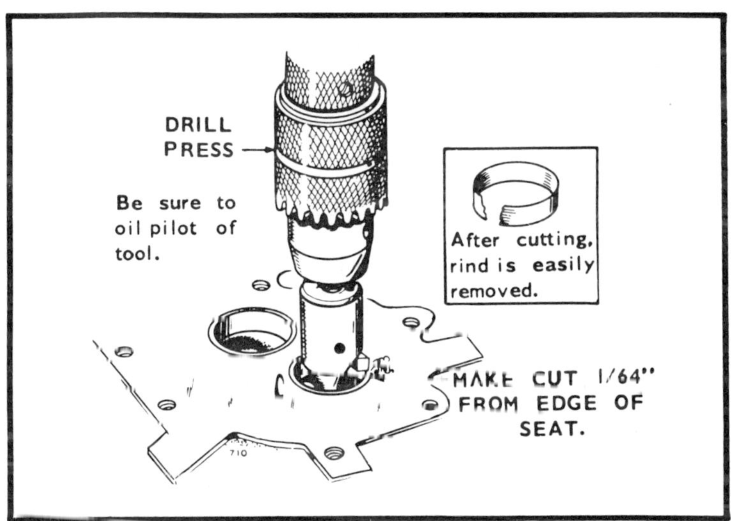

Fig. 6-12. Cutting out valve seats is a simple operation. (Courtesy Onan.)

In addition, valve seat and valve face angles are never quite identical. For example, Bedford intake and exhaust valve faces are 29 and 44 degrees, for a 1 degree interference with their seats. Waukesha VR intake valves are ground at 29° 30′ and the exhausts at 44° 30′, for a half-degree interference. Note that the odd angle is always on the valve face, never on the seat. It is easier to adjust a valve lathe for the required angle than to dress a seat stone for it.

Seats are machine-ground with a hand-held tool centered on the guide. Unless factory specs or a careful examination of the port shows otherwise, the inboard cut is with a 60 degree wheel and the outer cut is with a shallow 30 degree stone. The area between these cuts forms the valve seating surface. Its height above the cylinder head is extremely critical, for if the valve is sunk too deeply in the port, gas flow will be disturbed and the compression ratio for that cylinder will drop. If it rides too high, standing proud beyond specification, compression will be increased and there will be danger of valve-piston collision. The width of the seat is a variable specification, and the rebuilder does have some say in the matter. A thin seat increases unit pressure and promises a better seal, but too much of a good thing can lead to maintenance headaches as the narrow seat pounds and flattens. The general practice is to make the exhaust valve seats wider than the intakes to improve cooling. The seat is the thermal high road from the valve head.

Valve Faces

Assuming that the guide is within specs and that the seat has been ground correctly, the valve will seal if these three requirements are all met:

1. The stem diameter is within the wear limit.
2. The stem is straight and at 90 degrees to the crown.
3. The face is ground at the correct angle.

Stem diameter can be determined with a micrometer and then compared against published factory specs, but stem straightness is best determined with a pair of ground V-blocks and a dial indicator. There are other ways that are less accurate but more practical for the amateur. You can chuck the valve in a drill press or a lathe (portable drill motors usually have bent spindles and are misleading) and watch for wobble. This will at least detect gross misalignments. The valve face grinder will have the last word on stem/crown concentricity. If the reground face is skewed to one side, the valve should be replaced.

Incidentally, some engines, particularly those with factory turbocharger installations, use sodium-filled exhaust valves. While

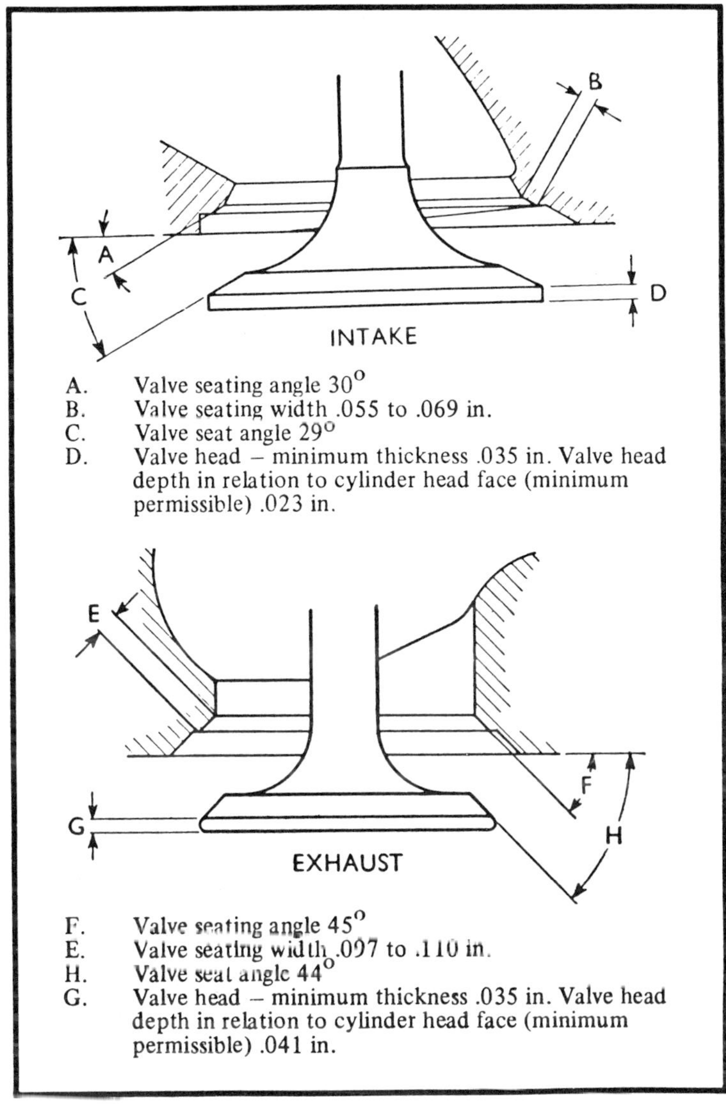

Fig. 6-13. Valve seating specifications for current Bedford production.

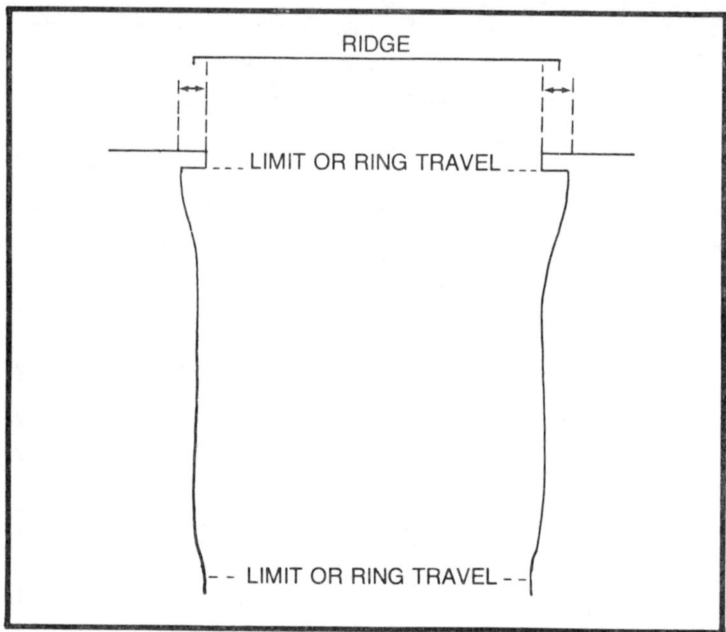

Fig. 6-14. Cylinder bore wear is, for practical purposes, contained within the area of ring travel. Distortion, on the other hand, involves the whole length of the cylinder.

there is no reason for anyone to saw a valve apart or splash a welding torch flame over it, either operation would be extremely hazardous. Sodium is unstable, combining explosively with water, even the water in your skin.

Once set up, the valves will stay gas-tight if the margin is within or greater than the specification, and if the valve springs are up to the job. The margin, dimensions D and G in Fig. 6-13, must be wide enough to stay relatively cool. The exhaust valve margin is the hottest spot in the engine and should be thicker than the intake. Specifications in Fig. 6-13 apply to a particular Bedford series and are not to be taken literally for other engines.

ENGINE BLOCK—INITIAL DISASSEMBLY

Cylinder bores wear in the pattern shown in Fig. 6-14. The lowest part of the bore, being remote from combustion heat and saturated with oil, receives little or no wear. Wear begins in the zone of ring travel and remains fairly stable until the upper limit of ring travel is reached. At this point the cylinder walls splay outward with the greatest loss of metal at 90 degrees to the piston pin. There are several causes for this localized wear: ring pressure, dilution of the lube oil with fuel, and acid attack. Of these, acid attack may be the most serious. The area above the first compression ring is unworn and forms the ledge which gives a rough-and-ready idea of cylinder bore wear: the more pronounced the ledge, the greater the amount of wear. But ledge width is only an approximation, and no substitute for precision measurement.

The ridge must be removed before the piston is withdrawn; otherwise the rings will snag on the ridge and hang, preventing piston extraction.

Figure 6-15 shows a ridge reamer, not one of the best, but sufficient for the amateur. The reamer should ride squarely on the block, with the cutter blade firmly against the wall. Turn counterclockwise, bearing down on the tool to keep it from climbing out of the bore. Cutter pressure is critical. Too much will distort the tool and leave a wavy cut; too little will allow the blade to skate and dull. With a little practice, you can sense the correct pressure by the swoosh of the blade cutting cleanly.

How much of the ridge should be taken out? In theory, all of it. In practice, the bore will be distorted and most worn in those areas that are perpendicular to the wrist pin. To remove all of the ridge would mean going deep, so deep that the upper ring would ride

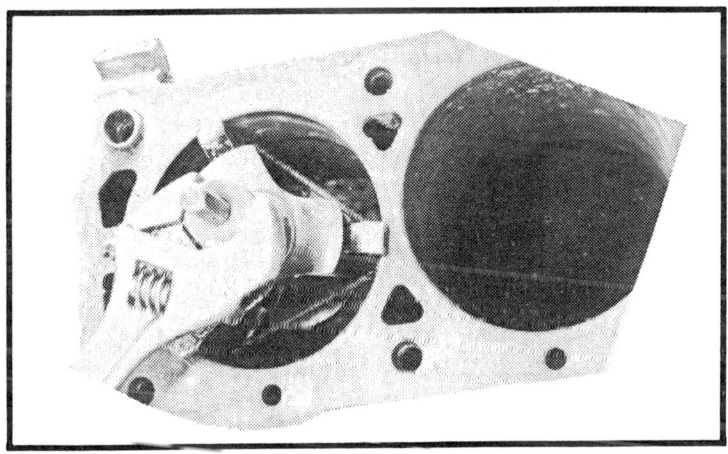

Fig. 6-15. Using a ridge reamer.

against the taper left by the reamer. At top dead center the ring would flex open and, a few hundredths of a second later, squeeze shut as the piston descends. The ring would quickly fatigue and snap. A little ridge is better than a broken ring, so stop when the ridge has disappeared from part of the bore.

Pistons

Remove the oil pan, noting the position of side-gaskets and end-seals. On some engines the sealing strips at each end of the pan are over the ends of the seals; on others, the seals are over the gaskets. Turn the crankshaft to bring No. 1 piston assembly to bottom dead center. Remove the connecting rod nuts or bolts.

Tap the rod cap to break the vacuum seal between it and the crankshaft journal. Lift the cap off, and with a hammer handle drive the piston and rod shank up out of the bore.

Observe that the rod and cap are stamped with the numeral 1 and that each carries embossed match marks that must be aligned in assembly. In addition, the piston/rod assembly must be installed so that the match marks are on the side of the engine as originally found. Otherwise, the rod's oiling system will be disrupted.

Once the assembly is out of the bore, install the cap on the rod, align the match marks, and lightly run down the nuts. While parts could be dumped loose into the wash bucket and assembled later by number, experience shows that keeping caps and rods together eliminates error.

Clean the piston tops with a wire wheel. It is neither necessary nor desirable to buff varnish from the sides of the piston; that which you remove will quickly form again, and once there will become some small benefit to compression. Clean the ring grooves with the appropriate tool which is available from auto supply houses. An occasional mechanic does get by with a broken ring held in a file handle, but the work goes slowly.

Erosion is possibly the most common form of piston damage The crown appears to have been nibbled by mice. Even though the marks may look superficial, it is better to replace a piston with this sort of damage. The cause is detonation, a condition brought on by one or more of the following:

- Habitual engine lugging. In this case the mechanic is home free, for he has only to admonish the driver to keep the revs up.
- Dribbling injectors. This condition will be caught by the injector specialist.
- Improper timing. It is almost always early.

Broken ring lands are the result of detonation or of:

- Excessive ring-to-groove side clearance, which is discussed below.
- Heavy use of starter fluid.
- Coolant leak into the cylinder. So check head and wet-sleeve gaskets.
- Failure to ream away the ridge before the piston is removed. In this case the damage will be under the ring.

Severe abrasion or scuffing on the thrust side of the piston, which are those areas prependicular to the wrist pin, is caused by:

- Insufficient bore lubrication. Check the oil pump, pickup screen, and crankshaft bearing clearances.
- Lugging.

Scores on the maximum and minimum thrust faces, which are those areas adjacent to the wrist pin, may be the result of the causes listed above, or may be the result of a piston that is too large for the bore. Scores confined to 45 degrees on each side of the pin, leaving the thrust faces unmarred, can almost always be traced to improper pin fit.

A runaway lock ring takes the end of the pin boss with it and may leave a well-marked trail on the cylinder bore. Expect the connecting rod to be bent. Lock ring problems can be traced to any of the maladies listed below:

- Taper at the crankpin journal that puts side-force on the rod.
- A bent connecting rod.
- Extreme crankshaft endplay.
- Overstressing on the lock ring on installation.

Modern pistons are cam-ground, which means that pistons are egg-shaped when cold (Fig. 6-16A). Only the major thrust faces contact the bore. At operating temperatures, the piston expands to

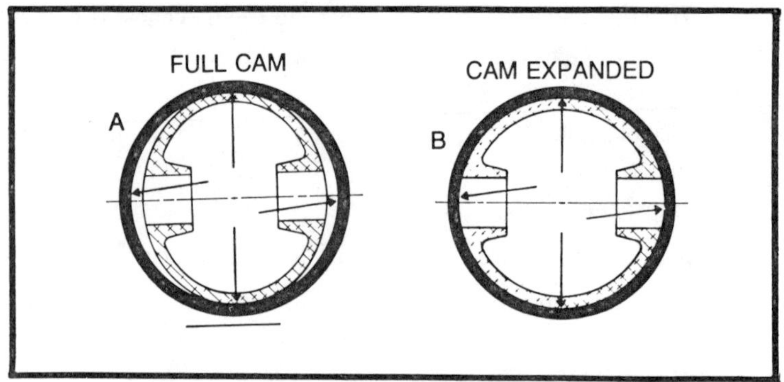

Fig. 6-16. Aluminum pistons are cam-ground to allow room for heat expansion. When cold the piston thrust faces touch the bore in what engineers call the full cam position (view A). As the engine warms, the piston fills the bore in the expanded cam position (view B). (Courtesy Sunnen Products Co.)

fill the bore (Fig. 6-16B). In addition, most pistons have a slight taper, narrowing toward the crown. Since pistons are miked cold, the readings must be taken between the maximum thrust faces approximately on level with the wrist pin.

Subtracting the piston diameter from the bore diameter gives the running clearance, one of the most critical specifications of the engine. Consult your dealer or the piston manufacturer for the exact data. Big Mack engines are set up so loosely that you can rattle the pistons in their bores by hand; clearance for Waukesha VRD 232 engines ranges between 0.0008 and 0.0028 in.

Wear patterns can be instructive. Normally, we expect uniform wear bands extending up the piston flanks and through the maximum-thrust faces. A bent connecting rod oscillates the piston and leaves hour-glass shaped wear patterns (Fig. 6-17A). If all pistons in the set show these patterns, figure the crankshaft is bent. A twisted connecting rod concentrates wear above and below the wrist pin centerline (Fig. 6-17B). Obviously, these faults must be corrected if the engine is to live.

Check the width of the ring groove with a feeler gauge and a *new* ring (Fig. 6-18). A narrower-than-specified groove means a parts mix-up: either the pistons or the rings are not mated. Too wide a groove results from wear and will be exaggerated by detonation, lugging, and high piston velocities. In any event, out-of-spec grooves are intolerable and will lead to ring breakage. Grooves can be machined oversize and sleeved down with spacers available from Sealed Power and other manufacturers. Most mechanics replace the pistons for the insurance value.

Unless a piston or rod is damaged, most mechanics leave tne parts assembled during the usual overhaul. Some idea of the pin fit can be had by holding the rod shank in one hand and moving the piston up and down with the other. A rebuild is a more thorough operation: each part must be disassembled and gauged.

Automotive engines have full-floating wrist pins. The pin pivots on the piston and on the small end, or eye, of the connecting rod. At room temperature, only the rod bushing has clearance; the piston bosses shrink and grip the pin with a force that requires several tons to overcome. In addition, the pin is located by a snapring at each end.

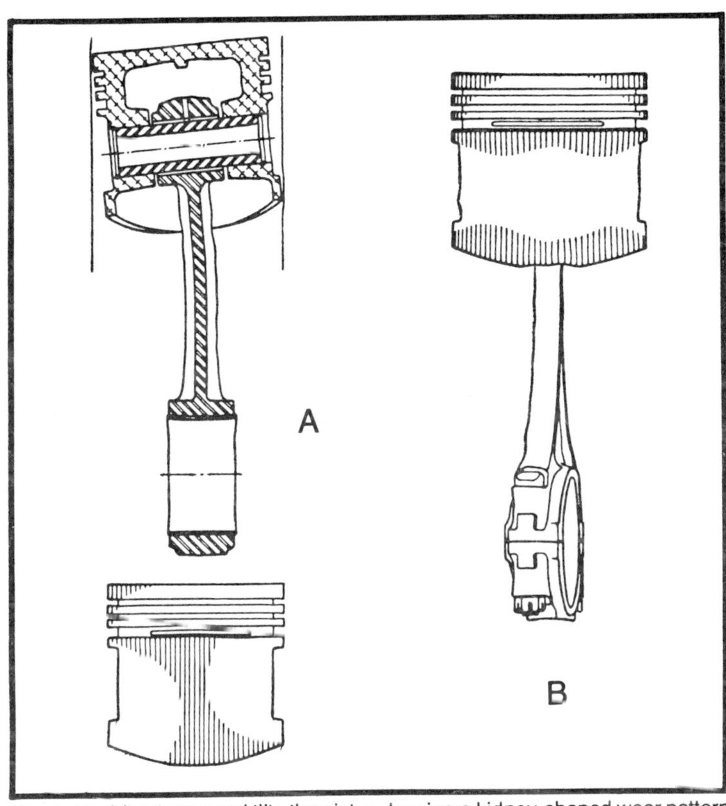

Fig. 6-17. A bent conn rod tilts the piston, leaving a kidney-shaped wear pattern (view A). A twisted rod rocks the piston, generating wear above and below the wrist pin (view B). (Courtesy Sealed Power Corp.)

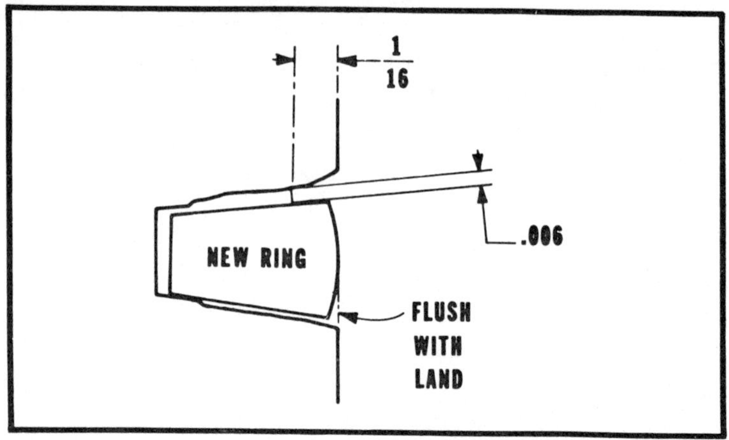

Fig. 6-18. One manufacturer's approach to the problem of measuring tapered ring grooves. (Courtesy Sealed Power Corp.)

Examine the piston and rod before disassembly. If the rod has no obvious reference points such as the oil bleed hole in Fig. 6-19, mark which edge of the rod is forward. Using a suitable tool, compress and withdraw the snaprings. Discard them. The snapring function is too critical to be entrusted to used parts. Mount the piston over an alignment fixture and press the pin out as shown in Fig. 6-20. If you do not have access to an arbor press, the piston can be heated with oil-soaked rags or with an electric hotplate. The pin should slide out easily.

Carefully inspect the rod and piston bearing surfaces. The pin should have full contact at each support, which will be indicated by dark areas on the rod bushing and piston bosses. Bright spots on the bronze bushing and on the aluminum bosses mean excessive clearance. As a rule of thunb, subject to correction by factory specs, the pin should have 0.0001-0.0003 in. clearance at the piston and 0.0003-0.0005 in. clearance at the rod. Increase rod clearance to 0.0005-0.0007 in. if the rod is drilled for full-pressure lubrication.

To assemble, align the piston and rod and lubricate the bearing surfaces with motor oil. Supporting the piston as before, press the pin home. If the piston is heated, the pin should enter with slight palm pressure and continue through to the far side of the piston without hesitation. If the pin pauses or clicks as it contacts the second boss, the pin bore is out of alignment.

Piston Rings

Although multipurpose rings have blurred the categories of late, there are three basic ring types to suit three different positions and purposes. The uppermost ring is always a compression ring, whose only function is to make a gas-tight seal against the bore. The second ring may be a pure compression ring, or it may double as a seal and a scraper. As a scraper its function is to squeegee surplus oil off the bore, so that the combustion chamber remains relatively dry. The oil control ring is always at the bottom of the stack. Some

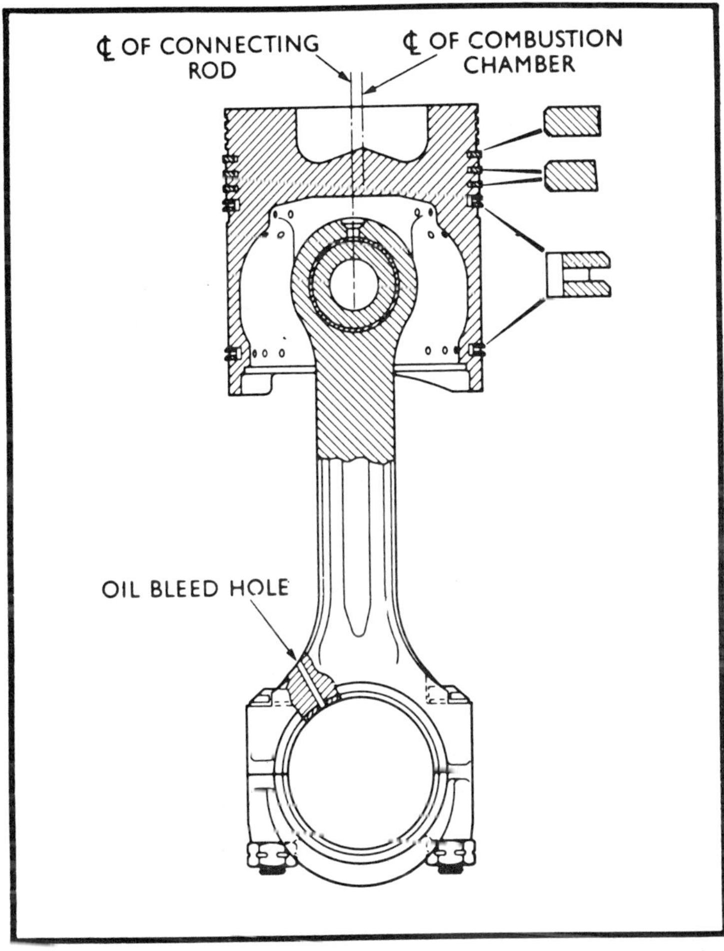

Fig. 6-19. A GM-Bedford piston and rod assembly. Note the piston offset relative to the centerline of the bore.

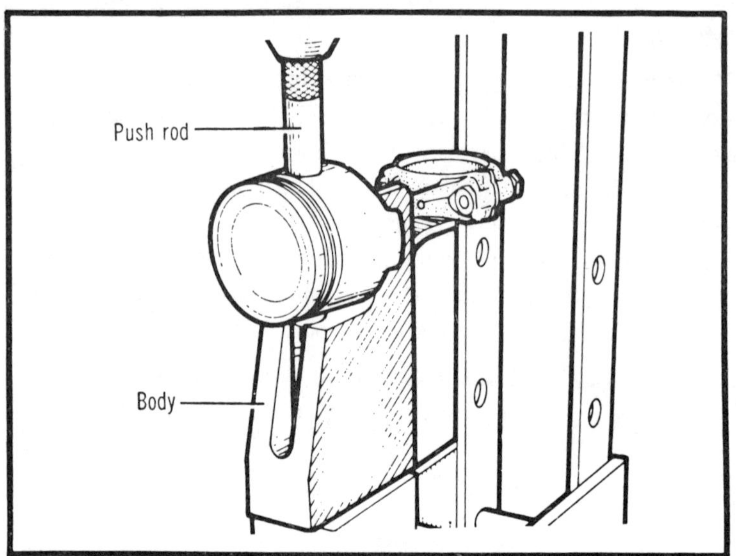

Fig. 6-20. Floating wrist pins are adrift only when the engine is warm; cold, the pins must be pressed out.

engines have two oil rings, the second one being under the wrist pin. These rings may be cast as a single part or may consist of upper and lower steel rails and an expanded steel element. In any event, the center section of the ring is relieved to distribute oil on the bore. The ring groove is slotted or drilled so that surplus oil can bleed back through the piston and into the sump.

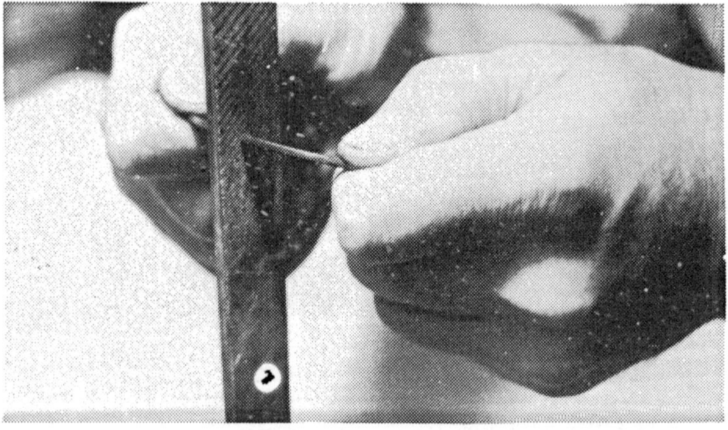

Fig. 6-21. Filing ring ends is tedious, but can save embarrassment later.

Once you are satisfied that the grooves are up to the job of locating the rings, determine the gap clearance. Using the head of a piston as a pilot, insert each compression ring about halfway down in the bore it will serve. Measure the gap with a feeler gauge.

Some clearance between the ring ends is needed to allow for heat expansion, but too much clearance encourages blowby and power loss. The traditional clearance was 0.003 in. per inch of bore diameter, but new ring alloys have made this formula obsolete. These alloys, known in the industry as "high strength iron," have different expansion characteristics and no single specification applies. Follow the ring manufacturer's recommendations.

Excessive clearance means that the bore has been worn out of countenance or, much less likely, that the wrong rings have been supplied. Too little clearance can be corrected with a file as shown in Fig. 6-21. Hold the ring flat against the file so that the edge is cut square. Confine your efforts to one end of the ring and check the progress in the cylinder bore.

With the help of a ring expander, mount the rings on the pistons (Fig. 6-22). Start with the bottom oil ring and work your way up. Oil

Fig. 6-22 Using a ring expander. On this particular piston the arrow points forward.

rings are symmetrical, without a definite top or bottom. However, scraper and compression rings have the upper side marked with a symbol or letter and installing these unidirectional rings upside down costs power and oil consumption.

Stagger the ring gaps per specification. If the factory manual is silent on this subject, arrange the gaps at 120 degree intervals on three-ring assemblies and at 90 degrees on four-ring groups. The idea is to block blowby during startup. Once the engine starts, the rings will rotate on the piston and the gaps will distribute themselves at random.

Install a new insert bearing half into the rod shank, aligning the tab on the insert with the groove on the rod. Oil liberally and flood the crankpin with oil. Dip the piston wrist-pin deep in a can of lube oil and allow a few minutes for the surplus to drain off. Mount a ring compressor over the piston, tightening it firmly enough to squeeze the rings into their grooves. For the insurance valve, cover the ends of the rod bolts with sections of fuel line or the bolt ends may otherwise nick the crankpin.

Place the piston over its cylinder bore with the front of the piston forward. Push the assembly down and out of the compressor. If tension is correct, you should be able to feed the piston into the bore with heavy thumb pressure. Stop if you feel the piston jar to a halt. This means that a ring has escaped the tool and continued pressure will damage the piston or break the ring. Once the piston has passed through the compressor, reach up from the underside of the engine and pull the rod down over the crankpin. Remove the fuel lines and check that the upper insert is in place. The rod shank number must be on the same side of the engine as it was originally.

Place a new insert in the cap for the rod. Oil it liberally, smearing the oil over the whole bearing surface with a clean finger, and mount the cap. Match marks must align. Torque in several increments to specification. Once the cap is secure, bar the crankshaft over to detect binds or hard spots.

Connecting Rods

A thorough job means that the connecting rods should be inspected for big-end stretch, shank distortion, and fatigue cracks. I will describe these procedures for whatever interest they may have

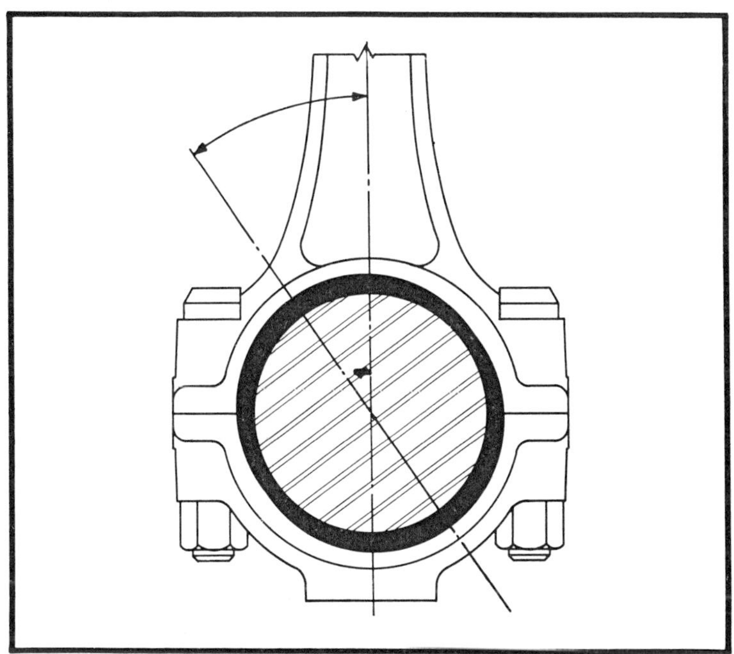

Fig. 6-23. Big-end stretch is confined to within 35 degrees of the connecting rod centerline. (Courtesy Sunnen Products Co.)

to the reader; however, most of this work belongs in the province of the automotive machinist who can recognize the faults and who has the equipment to correct them.

Unless the bearing has wiped, big-end stretch is confined to 35 degrees on each side of the vertical (Fig. 6-23). Engineers cannot say why it occurs, but it is a fact of life for high-speed diesel and gasoline engines. The effect is to increase bearing clearance and to destroy the uniformity of the oil cushion between the insert bearings and the crankpin. The bearing is liable to seize. Even if it doesn't, excessive oil will be thrown off the crankpin and may find its way into the combustion chamber.

Reconditioning involves:

- Measuring the big-end bore to determine the amount of stretch (Fig. 6-24A).
- Jig-grinding the rod and shank faces (Fig. 6-24B).
- Assembling and torquing the rod cap (Fig. 6-24C).
- Honing the bore to the original diameter (Fig. 6-24D).

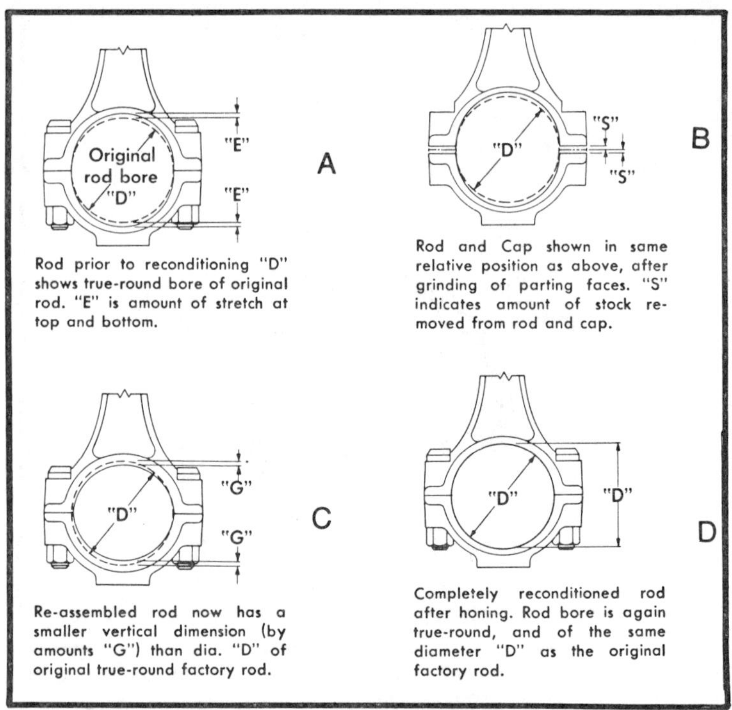

Fig. 6-24. Reconditioning a rod big-end involves grinding the parting faces and honing the bore. (Courtesy Sunnen Products Co.)

The metal removed does not seem to have any adverse effect upon rod strength; for practical purposes, the rod is as good as new.

Fatigue failure is fairly common on rods that have seen much service. The rod grows brittle from flexing and then shatters like glass. Since fatigue cracks grow slowly and from the surface inward, most of these failures can be prevented by early detection. The catch is that most cracks are invisible to the naked eye and must be amplified by artificial means. The most widely used technique is the Magnaflux, or magnetic particle method. The part under inspection is sprinkled with iron filings and charged with a heavy direct current of 300 amps or more. Surface flaws become magnetic poles and attract the powder. Cracks that were invisible before become valleys between ridges of iron filings.

All connecting rods, even brand new ones, have surface defects (Fig. 6-25). The trick is to distinguish between foundry marks and cracks that go to the heart of the rod.

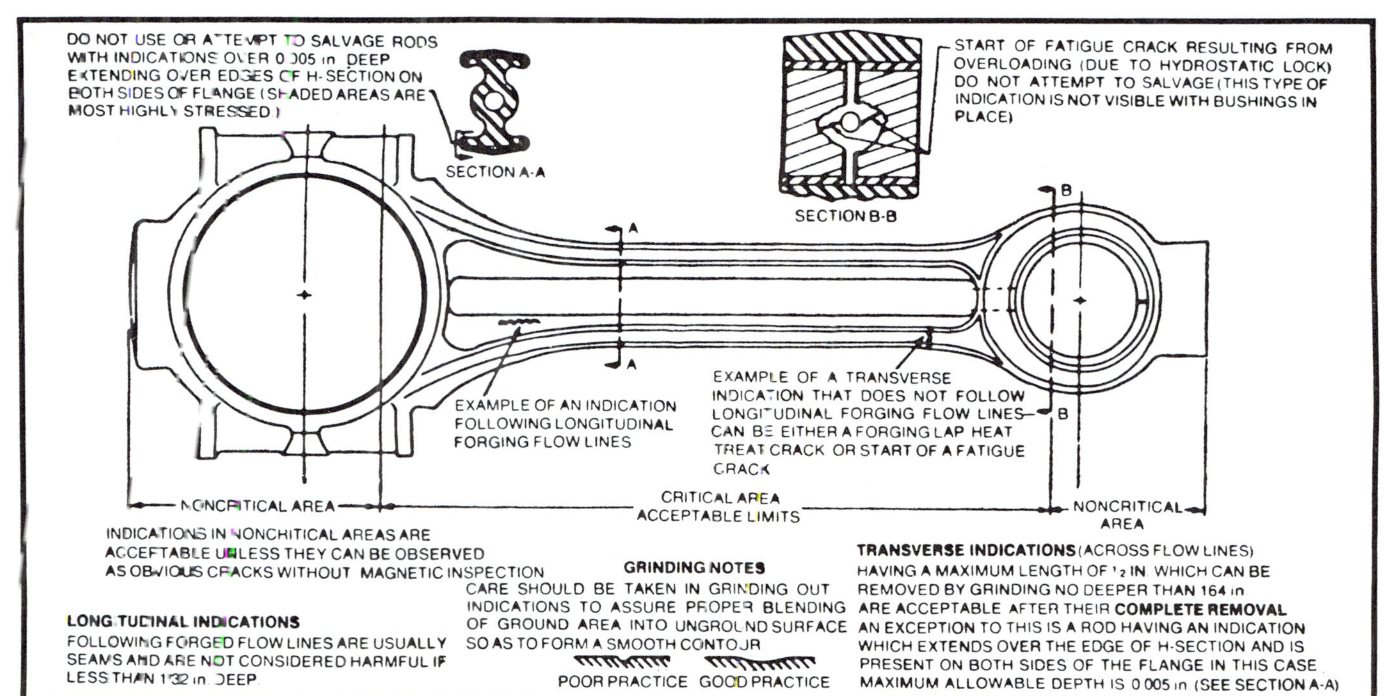

Fig. 6-25. Skill is required in interpreting Magniflux readings since all connecting rods carry some flaws. (Courtesy Detroit Diesel Allison.)

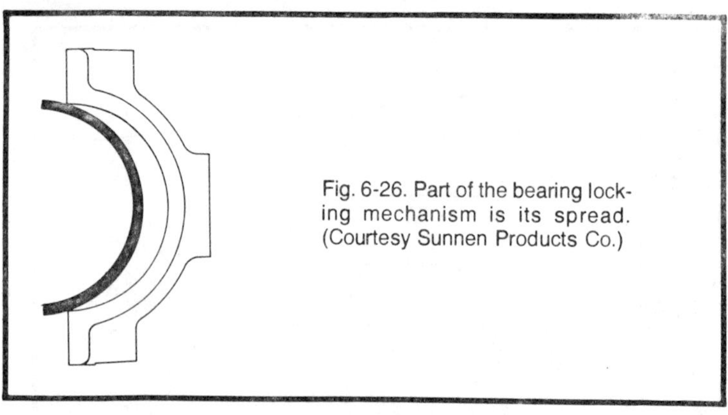

Fig. 6-26. Part of the bearing locking mechanism is its spread. (Courtesy Sunnen Products Co.)

In general, longitudinal cracks are harmless unless they are 1/32 in. deep or more. Transverse cracks, cracks running across the width of the rod, are dangerous. If they are more than 0.005 in. deep and run over the edges of the H-section, discard the rod. Transverse cracks confined to the sides of the rod that are less than 1/2 in. long and 1/64 in. deep can be ground and feathered. Cracks on the small end are the result of improper pin fit or of impact damage, and mean that the rod should be replaced.

Insert (Rod and Main) Bearings

Insert bearings have these critical dimensions:

- The spread which makes it necessary to snap the insert into the rod or cap and holds the insert in place during assembly (Fig. 6-26).
- The crush, which is the distance each insert half stands proud, above the parting line (Fig. 6-27). The purpose of crush is to force the inserts tightly against the rod bore after assembly. It is an important enough measurement for diesel manufacturers to have published a specification.
- The wall thickness which is the variable that distinguishes between a standard and an undersized bearing (Fig. 6-28).
- The insert length, which is of more concern to the bearing manufacturer, but will cause the mechanic trouble if the length is excessive and the bearing edges ride against the crankpin chamfer.

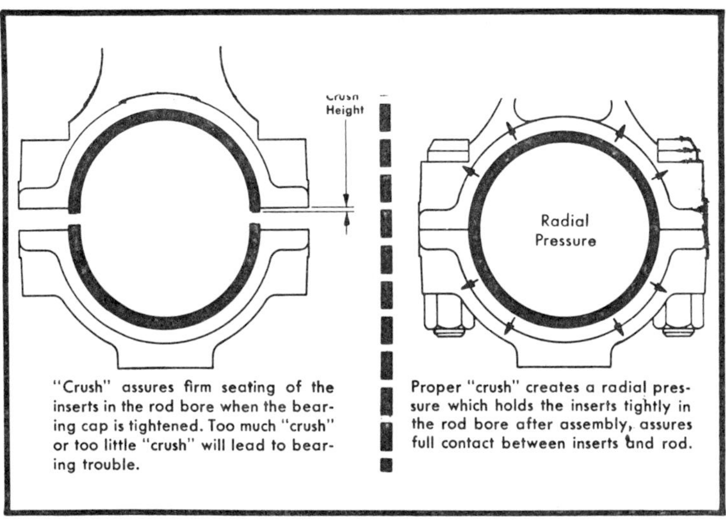

Fig. 6-27. Bearing crush is critical and many engine manufacturers supply specifications for it. (Courtesy Sunnen Products Co.)

- The inner diameter of the insert, which is of prime consideration to the engine rebuilder. The difference between bearing inner diameter and crankpin outside diameter is the bearing clearance.

There are two ways to measure bearing clearance. The traditional method is to measure the bearing's inner diameter with an inside micrometer and subtract this reading from the crankpin mea-

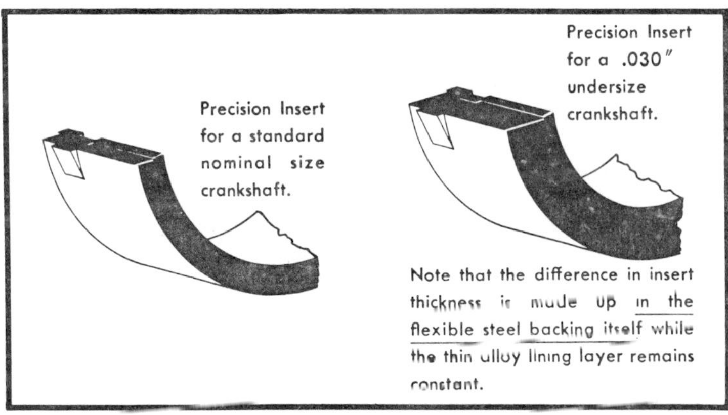

Fig. 6-28. Undersized inserts have thicker backing. (Courtesy Sunnen Products Co.)

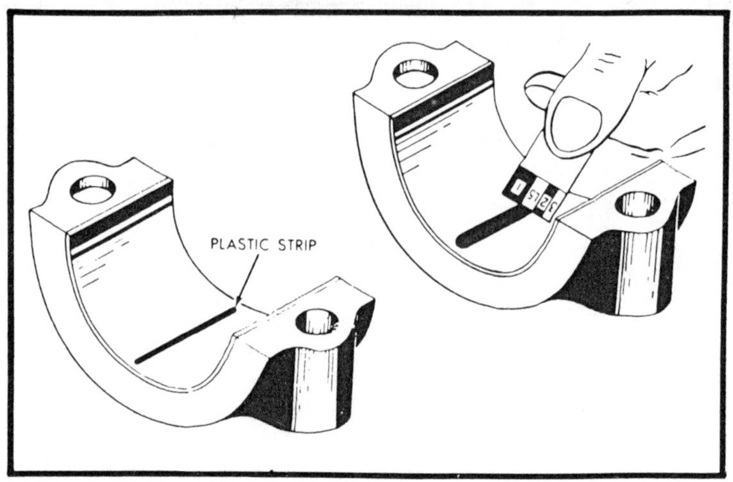

Fig. 6-29. Using a plastic gauge. (Courtesy Detroit Diesel Allison.)

surement. While this method is accurate, it requires two precision gauges and the skill to use them. Many mechanics have gone to the plastic gauge method, which gives results that are as accurate as those given by micrometers, but which involves an investment of 35¢ and less patience.

The gauge material is a soft plastic strip, color-coded by size. For example one manufacturer uses a green strip for the narrow clearances expected of new bearings, and blue for the greater clearances associated with worn bearings. Both the crankpins and bearings should be dry because the plastic will dissolve in oil. Pinch off a strip of it and lay it over the center of the cap insert as shown in Fig. 6-29. Assemble the cap on the rod, aligning the match marks, and torquing to specification. Be careful not to move the crankshaft in the process. Lift the cap and read the bearing clearance with the help of the scale on the gauge package. A difference in the width of the gauge represents crankpin taper. Some idea of crankshaft eccentricity, or "out-of-round," can be had by positioning the gauge lengthwise in the cap. Scrape off the flattened gauge material before final assembly.

Connecting Rod Assembly

Flood the inserts and crankpin with clean lube oil, making sure that all rubbing surfaces are wetted. Lubricate the rod bolt threads and install the cap, with the match marks aligned. Pull the cap bolts down evenly and torque to specification.

Camshaft Drive

Camshaft, high-pressure pump, and, as the case may be, balance shafts drive off the crank. Power is transmitted by gears or by chain (Fig. 6-30). Before disassembly, align the marks so that you will understand their significance. It is traditional practice to index the marks with No. 1 piston at top dead center on the compression stroke, although some engines are timed differently. Figure 6-31 illustrates the arrangement used on Chrysler-Mitsubishi engines and Fig. 6-32 shows the Detroit-Diesel layout with provision for right- or left-hand crank rotation. Align sprocket timing marks with a straightedge since the unaided eye can be deceived.

Wear in the timing gears can be detected with a dial indicator or indirectly with the help of a piece of solder. Place the solder at the base of a tooth on the gear set that is to be checked and bar the engine through. The width of the solder indicates gear lash. Timing chains are best replaced during overhaul, since even the best of them wears rapidly.

Crankshaft

Invert the cylinder block and detach the flywheel and engine pulley. If the crankshaft turns as you remove the bolts, wedge it against the block with a two-by-four.

Fig. 6-30. Like many other automotive plants, the Mercedes-Benz 190 employs a timing chain.

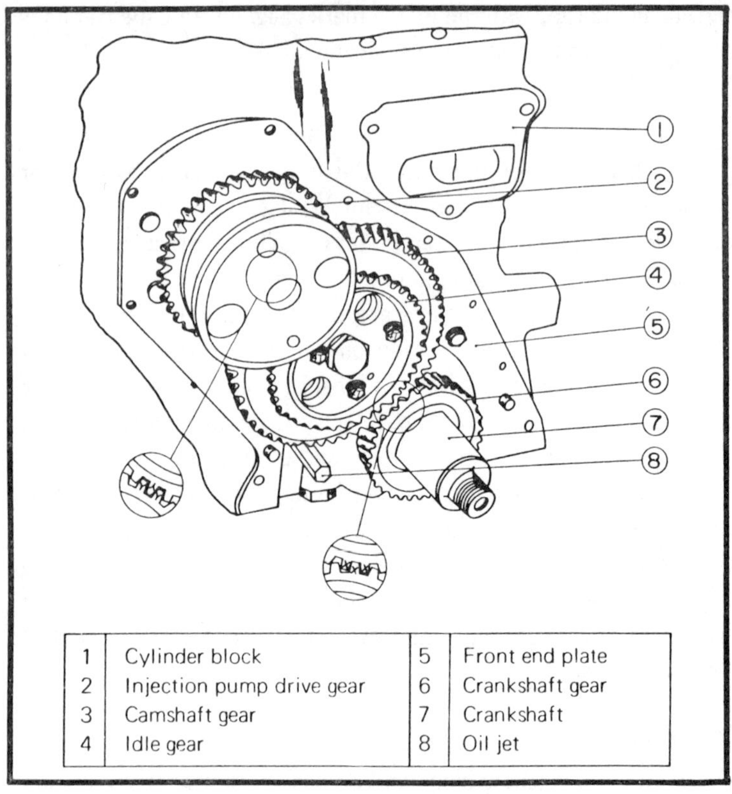

1	Cylinder block	5	Front end plate
2	Injection pump drive gear	6	Crankshaft gear
3	Camshaft gear	7	Crankshaft
4	Idle gear	8	Oil jet

Fig. 6-31. Mitsubishi timing marks.

The main bearing caps should be marked with an arrow that points to the front of the engine. In addition, each cap should be identified by number with No. 1 cap closest to the radiator; No. 2 next; and so on. These marks are usually present. If they are not, make them yourself. Sometimes an engine rebuilder will change the cap position and not indicate what he has done with new marks. Failing to report his work is inexcusable, but the work itself is permissible so long as the machinist line-bores the bearings. This process is discussed under the head "Main Bearings."

Inspection

An engine is only as good as its crankshaft. Because of the severe stresses upon them and because of the long hours they endure, diesel crankshafts are subject to catastrophic failure.

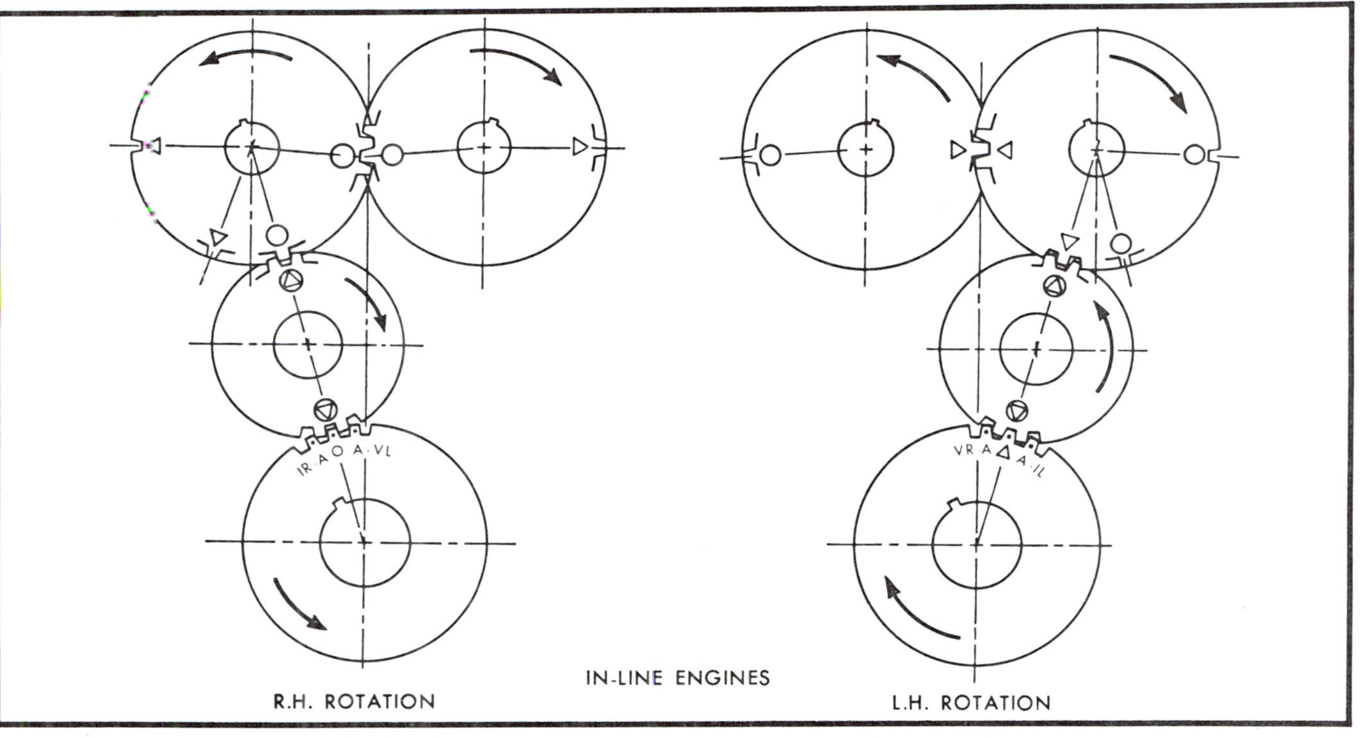

Fig. 6-32. Detroit engines can be set up for left- or right-hand rotation.

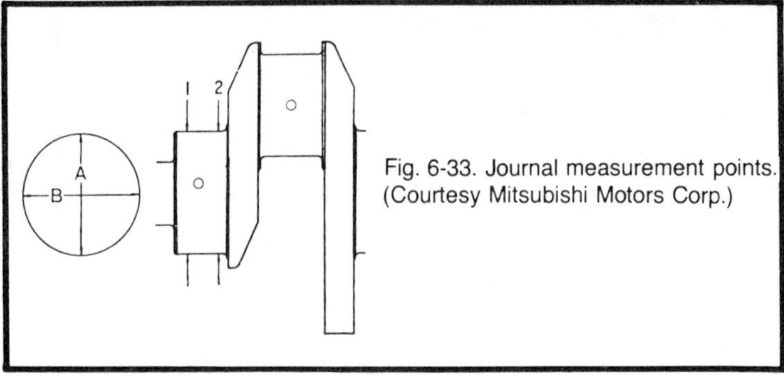

Fig. 6-33. Journal measurement points. (Courtesy Mitsubishi Motors Corp.)

Make this series of inspections:

- Have the crankshaft blow-tested before you begin to invest time or money in it.
- Examine the bearing journal for light ridging caused by oil distribution grooves in the bearing inserts. Polish out these imperfections with crocus cloth wetted in fuel oil. The crocus should be nearly as wide as the journal and slightly longer than its circumference. Loop the abrasive over the journal and turn with a leather thong. The thong should be crossed under the abrasive so that it applies the same pressure on all sides.
- Examine the oil-seal rubbing surfaces at the shaft ends. Sometimes these surfaces are renewable and at other times the front seal can be moved to bear against virgin metal. Otherwise the scores should be polished out as described above. Use a fairly coarse abrasive for this: mirror-like surfaces weep oil.
- Mike the bearing journals as shown in Fig. 6-33. Compare the diameter and eccentricity, or "out-of-round" taper, with factory wear limits.
- Check the main bearing thrust face, which is often but not always at the center main, for scoring. Light scores can be smoothed out with a stone; deeper marks will require the attention of a machinist.
- Have crankshaft trueness checked. If you are making a serious job of this project and have the money to spend, bring the piston/rod assemblies along and have the engine balanced.

- Ideally the drilled oil passages should be opened for cleaning. Threaded plugs are reusable; soft plugs can be removed with a screw extractor and should be replaced with new ones.
- If the oil ports have square edges where they emerge into the journals, break the edges with a hand-held grinder. A 1/32-in. fillet is more than enough.

Flaw Testing

Most shops use the magnetic particle method to detect fatigue cracks. As in the case of connecting rods, crack interpretation requires practice. Forces acting on the shaft are shown by arrows in Fig. 6-34. The most critical load paths are through the fillets (shaded in the drawing) that join the crankpins with the main bearing journals. Figure 6-35 illustrates the crack pattern these forces generate. Any of these cracks is serious and spells the doom of an otherwise good crankshaft.

Premature failure, which is anything before a quarter-million miles or so, can sometimes be traced to detonation, over-speeding, lugging, or to some mechanical fault like a loose flywheel or too much drive belt tension.

Machining

Bearings are available in several oversizes to compensate for crankshaft wear and machining operations. Standard American oversizes are 0.001, 0.002, 0.005, 0.010, 0.020, and 0.030 in. The first three oversizes can be used without preparatory grinding; 0.010 in. and greater oversizes require that the crank be ground.

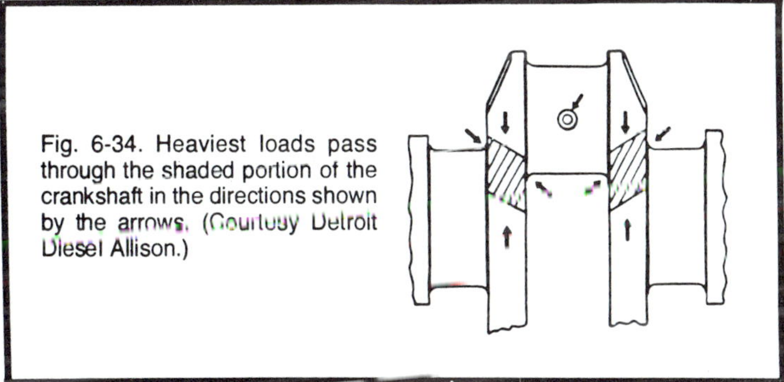

Fig. 6-34. Heaviest loads pass through the shaded portion of the crankshaft in the directions shown by the arrows. (Courtesy Detroit Diesel Allison.)

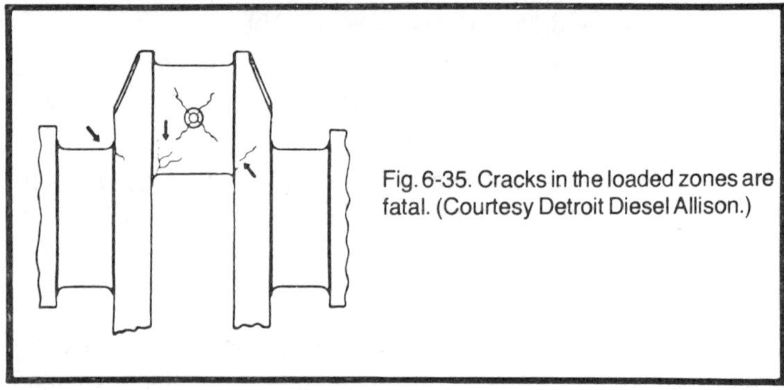

Fig. 6-35. Cracks in the loaded zones are fatal. (Courtesy Detroit Diesel Allison.)

Foreign engines, except those built in England, are set up around metric oversizes, the typical range being from 0.125 to 1.250 mm.

Although it is possible to grind a single journal or crankpin, this would be bad practice. *All* should be ground, even though only one has failed. Fillets, which are the curved rises at each end of the bearing and web, are critical. Irregular or "tight" fillets are stress rises. Wide, gently sloped fillets can bind the bearings.

The machinist should have the bearings in his possession; then bearing fit is his responsibility.

Note: Some International Harvester engines can come off the line with Elotherm crankshafts. These cranks are harder than conventional ones and can be ground only with great difficulty and under the supervision of IHC personnel. Attempts to straighten Elotherm crankshafts have been disastrous.

Main Bearings

The number of main bearings is an index of the strength of the lower engine. This index is by no means infallible: the Chrysler "Slant-Six" with three main bearings is one of the most durable auto engines. However, few builders would turbocharge an early Mercedes engine like the one shown in Fig. 6-30. Later M-B engines have five main bearings in the four-cylinder versions and seven in the five-cylinder model.

The upper halves of the bearings are integral with the block and get additional support from cast-in webs. Detachable caps form the lower halves of the bearing. While not universal, many diesel engines have deeply skirted blocks, extending well below the crankshaft centerline. The sides of the bearing caps are machined to make an interference fit with the block. Once the bolts are removed, the caps are lifted with a puller (Fig. 6-36).

The rear cap usually has some oil sealing function and should not be confused with the others. One cap, usually the center one, is machined on its trailing edge to accept the thrust bearing. All engines develop some crankshaft thrust, but automotive engines develop more in reaction to the clutch throwout bearing.

The remaining caps may appear interchangeable, but are not. As mentioned earlier, the bearing caps must be numbered by sequence and marked to indicate the front of the engine.

Main bearings are subject to misalignment, stretch, and spin. Some block warp is inevitable as the foundry stresses relieve themselves (Fig. 6-37). Since misalignment is a glacier-slow process unless the engine has been severely overheated, the original parts adjust themselves to it. When new insert bearings and a remachined

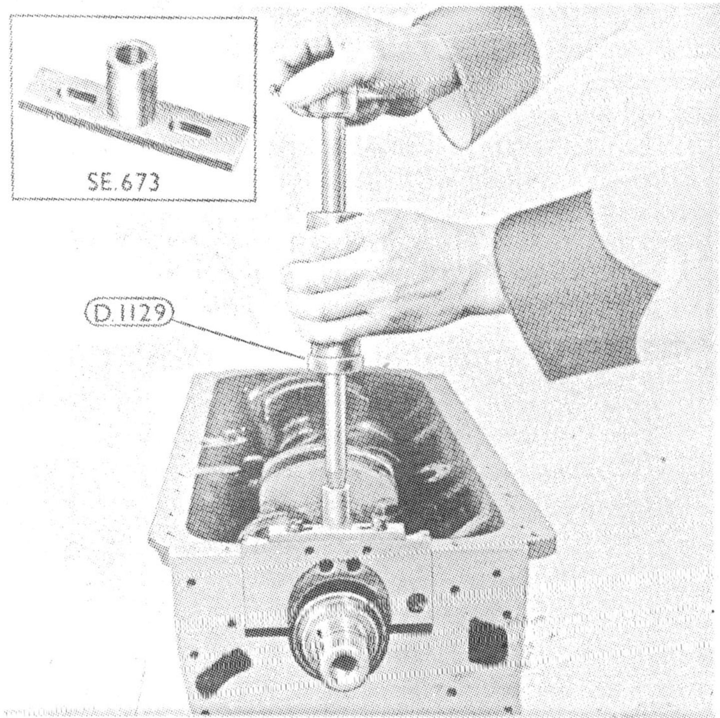

Fig. 6-36. A puller makes cap removal easier. (Courtesy GM-Bedford Diesel.)

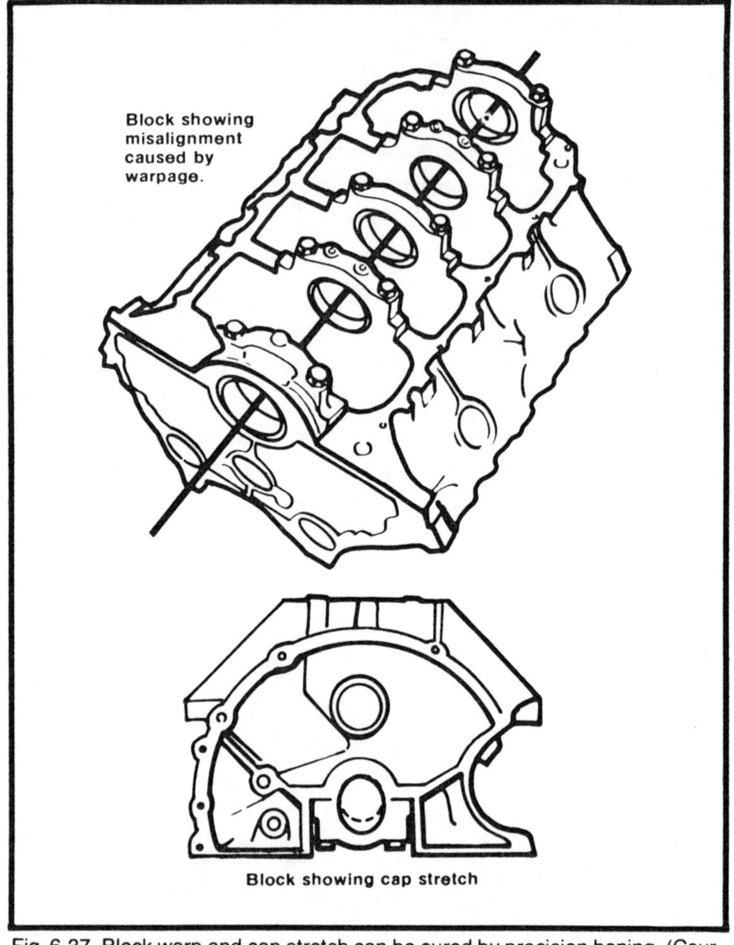

Fig. 6-37. Block warp and cap stretch can be cured by precision honing. (Courtesy Sunnen Products Co.)

crankshaft are installed, there is no compensation and the crankshaft will bind to some extent. It may even refuse to turn.

High loads can cause the caps to stretch out vertically and pinch in at the parting line as shown in the lower drawing in Fig. 6-37. As before, the original parts learn to live with this condition. New inserts will bind and can fail. Spin, in which the inserts break free of their locking tabs and turn in the bearing, is mercifully rare but does happen. Heat and load are the culprits.

The cure for all this is to line-bore or hone the bearings each time the engine is overhauled. In general a hone, like the one pictured in Fig. 6-38, is preferred to a boring bar. Each cap is torqued down to specification and the machine is passed through them. Since the bearings have already been honed at the factory, you can appreciate the importance of installing them correctly and in sequence.

Because of the conservatism of diesel enginemakers, rope seals continue to be used, but their days were numbered with the introduction of synthetic seals. A synthetic seal is dynamic in that oil pressure presses the seal more tightly against the crankshaft. Engine friction is reduced and, if the rubbing surface is regular, seal life will be extended.

In principle, the steep edge of the seal faces forward toward the bearing with the oil pressure leaking past it. In practice, seal contours are not always obvious to the eye. Read the instructions on the seal package carefully. If the crankshaft is out of the block, you can gain some small degree of insurance by dulling the bearing groove edges with a stone. It is easy for the seal to cut itself on these edges as it is snaked into place; so by all means, dull the cap edges.

Full circle seals that have neoprene sealing lips encased in a steel ring give better security than the two-piece variety. These seals are always used at the front of the engine and sometimes at the rear as well. Figure 6-39 shows two examples of IHC production in cutaway. One disadvantage of a full circle rear seal is painfully obvious to the mechanic who discovers he must pull the engine or transmission to replace the seal. On the other hand, these seals rarely fail.

Note that the IHC engine has a protective sleeve between the seal and crankshaft proper. After many hours the sleeve will notch and must be carefully chiseled off. The replacement is pressed on. The same arrangement is used on other industrial engines, front and rear.

The front seal is press-fitted into the timing case cover. Remove the cover and pry the old seal out with a screwdriver. By traditional practice, factory code numbers are stamped on the inner, or visible, side of the seal. The presence of these numbers means that this side of the retainer can withstand the pressure of installation. In any event, the steep edge of the lip should lie toward the sump. Most front seals bottom on a flange in the timing case. If the

Fig. 6-38. A main bearing hone.

flange is not present, some small change in seal depth is allowed to compensate for crankshaft wear. However, the seal must not stand proud of the case cover and must not be so deep that the drain port is obstructed.

Coat the outer diameter of the seal retainer with nonhardening adhesive, being careful to keep the adhesive off the seal lips. Press the seal home with a suitably sized driver and wipe off surplus adhesive. Oil the seal lips. Before installing the timing case cover, wrap masking tape over the crankshaft keyway. The keyway edges are sharp and could nick the seal.

CYLINDER BORES

There are three types of cylinder bore construction.

1. Wet sleeve. A sleeve, or liner, is inserted into the water jacket, gasketed, and secured against a flange by the cylinder head. This type of construction is used by Detroit Diesel and Peugeot among others.
2. Dry sleeve. The sleeve is pressed into the block where it is isolated from the coolant by block metal. This approach eliminates the danger of coolant leaks but reduces heat throw efficiency. Heat must pass through the sleeve and then across the barrier formed by the block boss before it is released into the coolant. Even though the sleeve has interference fit with the boss, the interface acts as a heat barrier. This quibble aside, most diesel engines are constructed with dry sleeves.
3. Sleeveless. The bore is machined directly onto the block, a lamentable practice which is followed by Bedford, Mercedes-Benz, and Oldsmobile. Sleeveless engines must be rebored when worn, for an increased displacement and compression ratio.

Inspection

Study the surface of the bore under a strong light. Deep vertical scratches indicate that the air filter failed or that the lube oil supply was contaminated. The causes of more serious damage, such as erosion from contact with fuel spray, galling from lack of lubrication, rips and tears from ring, ring land, or wristpin failure, will be painfully obvious.

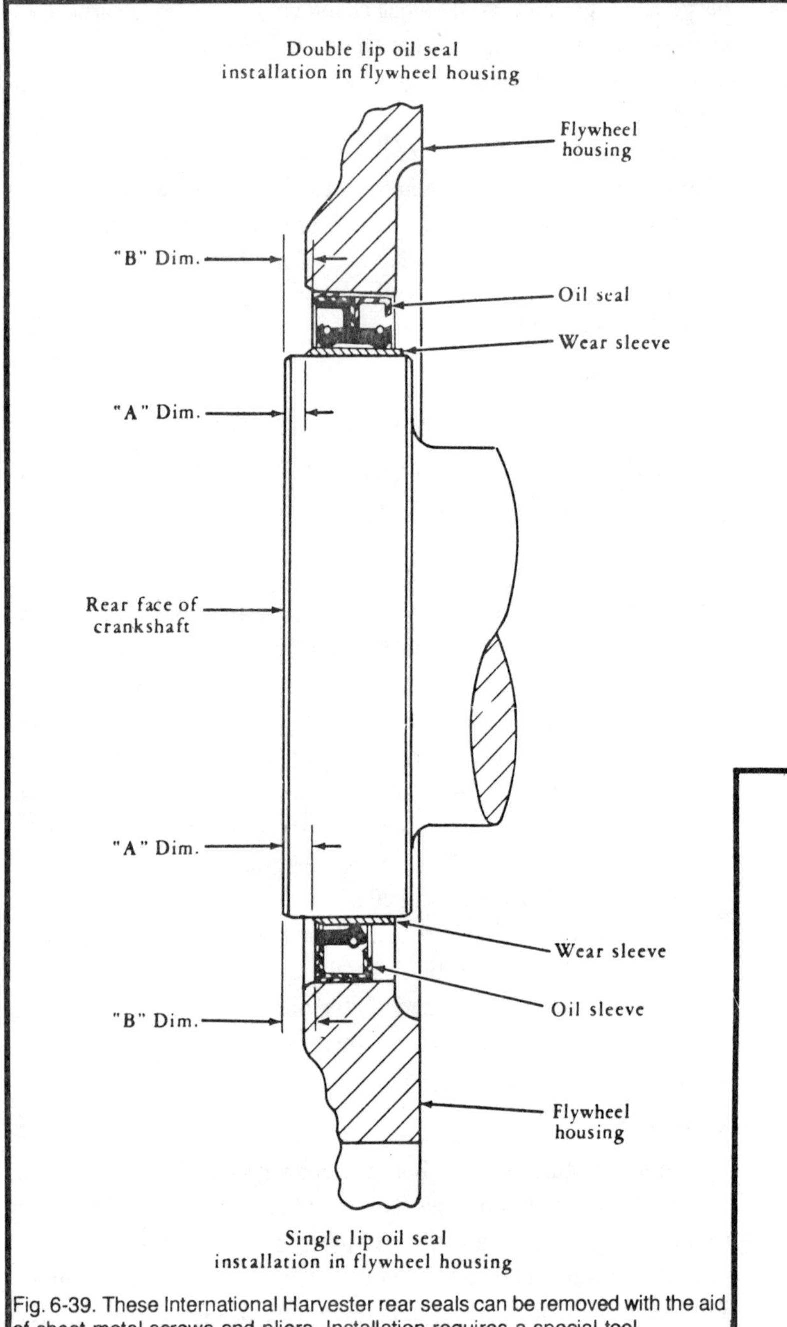

Fig. 6-39. These International Harvester rear seals can be removed with the aid of sheet metal screws and pliers. Installation requires a special tool.

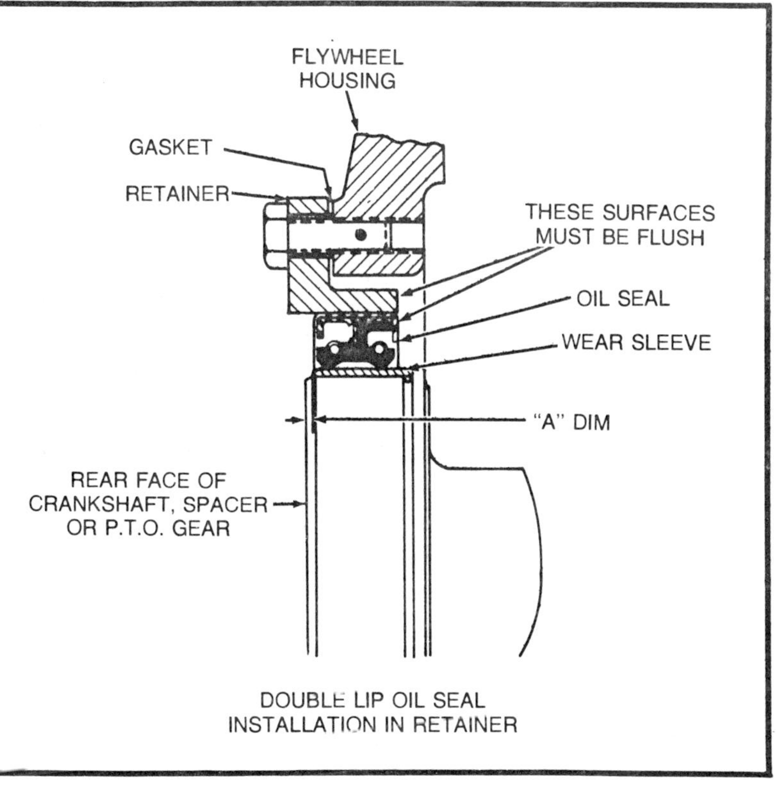

DOUBLE LIP OIL SEAL
INSTALLATION IN RETAINER

A cylinder gauge is the most accurate way to determine bore wear, taper, and eccentricity (Fig. 6-40), although you can use an inside micrometer. Measure at three places in the bore: the upper limit of ring travel, at midbore, and at the lower limit of travel. At each position measure bore diameter parallel to the crankshaft centerline and at 90 degrees to it. The second of these measurements will be the larger of the two, since it is taken across the bore thrust areas.

Boring

If the bore is worn out of spec, it will have to be machined oversize or, in the case of sleeved engines, replaced. The machining operation is generally done with a boring-bar, indexed to be at 90 degrees to the crankshaft centerline. While some boring-bar manufacturers assert that no finish honing is required, their claims are optimistic. Unless the machinery is new and the bit is perfectly

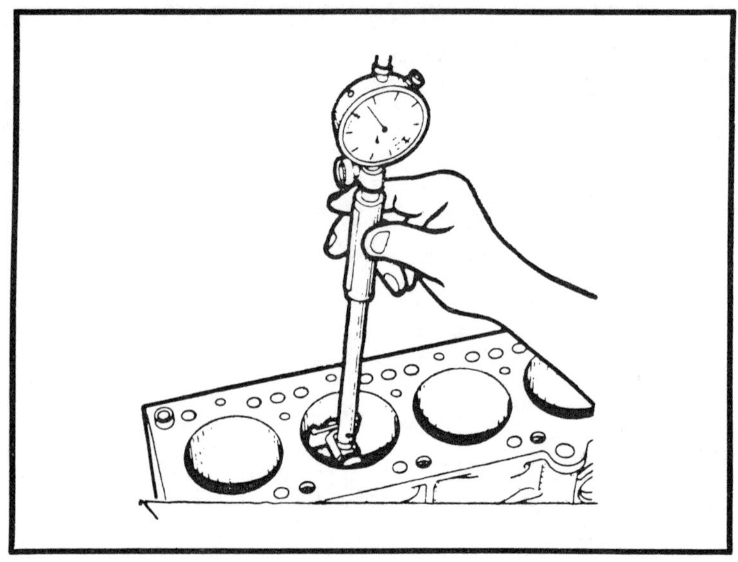

Fig. 6-40. A cylinder gauge is a convenience, although an inside micrometer can also be used to determine bore diameter. (Courtesy Chrysler Corp.)

sharpened, the bar will leave threads in the bore. These threads may not be obvious to the eye, but they do exist and form a very inhospitable surface for the piston and rings.

Honing and Glaze-Busting

A reputable machinist will leave 0.02 in. in the bore for the final honing operation. Since a significant amount of metal will be ground away, it is necessary to use a precision hone, one whose tension can be monitored, stabilized by a cylinder plate. These plates are made of heavy steel, sized to the engine at hand and torqued to the firedeck. This compensates for bore distortion caused by the cylinder head bolts. The main bearing caps are torqued for the same reason.

Glaze-busting is a much less formal process. An ordinary spring-loaded hone, powered by a drill motor, is adequate (Fig. 6-41). Used cylinder bores are covered with glaze, which is a kind of callus made of compressed and polished iron. Because glaze is such an excellent wearing surface, it must be removed so that the new rings will break in. Otherwise, the engine will continue to burn oil for thousands of miles, particularly when hard chrome rings are fitted.

Cutting stones are available in a variety of grits. The larger the grits, or stone particles, the harsher the action. Grit number, a term used by some manufacturers, is the number of stone particles that will cover one square inch. For example, a 280 grit stone is made up of particles that average 1/280 in. Other manufacturers prefer to grade their products by series number. A 280 stone corresponds to series 500.

As a general rule, the harder the rings, the finer the cut. Cast iron rings seat best with a 200 series stone; chrome rings require a 300 series stone; and stainless steel needs a 500 series.

The cross hatch pattern should look like the drawing in Fig. 6-42. Engine manufacturers disagree about the optimum angle, but something on the order of 30 degees from the horizontal is adequate. Steeper angles do not hold enough oil and shallower angles are too abrasive. A reciprocating speed of 70 strokes a minute with the drill motor turning a leisurely 400 rpm should do the job.

Keep the stones well lubricated with cutting oil or solvent and extend the stones a half-inch or so past the bore at the end of each stroke. Keep the hone moving in the bore. If you linger at the top and bottom of the strokes, the cylinder will bell-mouth.

Scratches are caused by dirty stones or insufficient lubricant. The stones can be cleared by shutting off the power and moving the

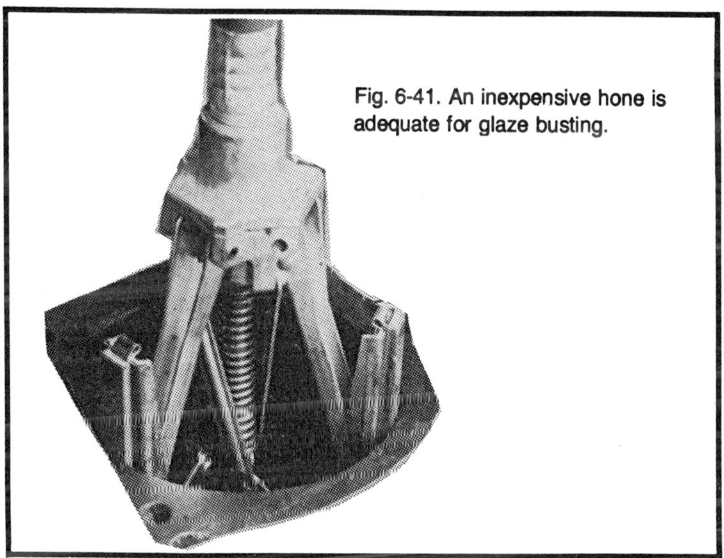

Fig. 6-41. An inexpensive hone is adequate for glaze busting.

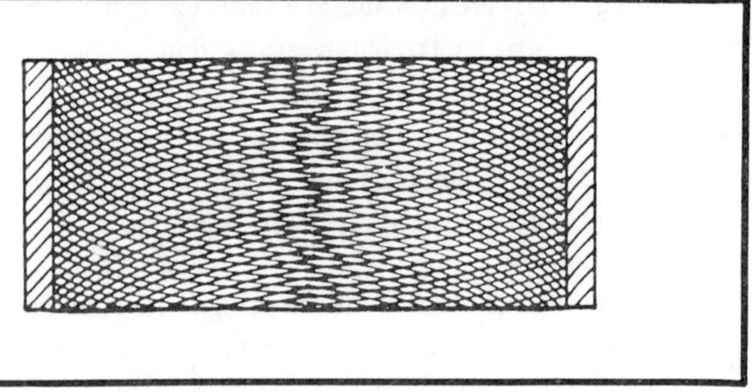

Fig. 6-42. Manipulated correctly, the hone will leave this pattern.

stationary hone up and down in the bore. Stop when the glaze is broken. Some worn or distorted cylinders may not clean up unless you remove large amounts of metal. Live with them: glaze is far less an inconvenience than an oversized bore.

Once the glaze is broken, remove all traces of abrasive. The only method that works is scrubbing the cylinders with detergent and hot water (Fig. 6-43). Solvent does nothing to remove the stone particles but will instead float them deeper into the crystalline structure of the metal.

Dry the bores with paper towels. If the towels stain, the bores are still dirty and must be scrubbed again. Oil immediately since clean cast-iron rusts within minutes.

RESLEEVING

Cylinder liners should be left in place during routine overhauls. What you can learn from removing them is not worth the effort and risk. Of course, if the block is to be boiled, wet liners and their composition ring gaskets must be removed, just as they must be removed if you suspect a water leak.

Wet liners will be gasket bound, but once this bond is broken the liners can be lifted out by hand. Dry liners that are pressed into place must be extracted with the appropriate tool. Figure 6-44 shows the business end of such a tool. The disc is positioned under the liner and then retracted with the help of a puller or slide hammer threaded into its boss. A force of several tons is sometimes necessary. If the liners are to be reused, identify each one by its cylinder number.

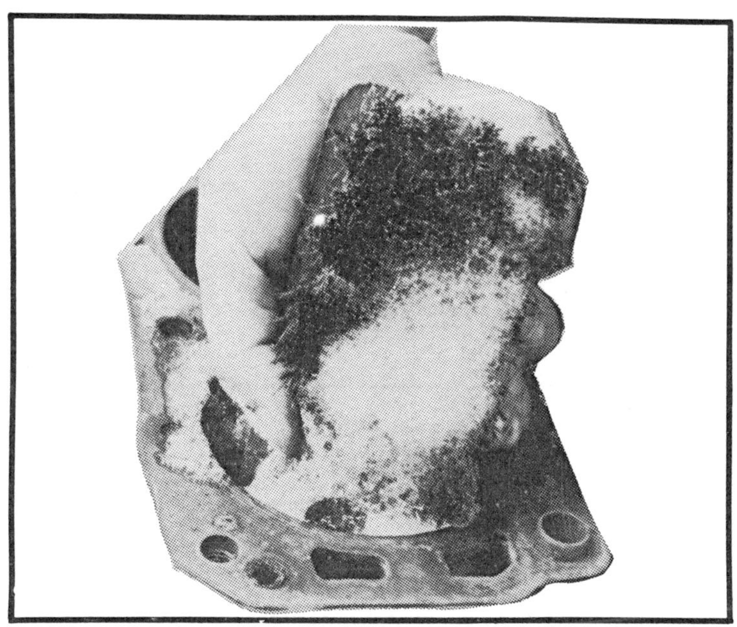

Fig. 6-43. Abrasive particles from the hone are stubborn and require detergent, hot water, and elbow grease to remove.

The critical aspects of wet liner installation are the sealing rings and the deck height, or the distance the liner stands proud of the block. Deck height determines the amount of crush supplied by the cylinder head. Too much crush will collapse the liner, too little will allow it to move and fret.

Fig. 6-44. A sleeve extractor for Mack engines.

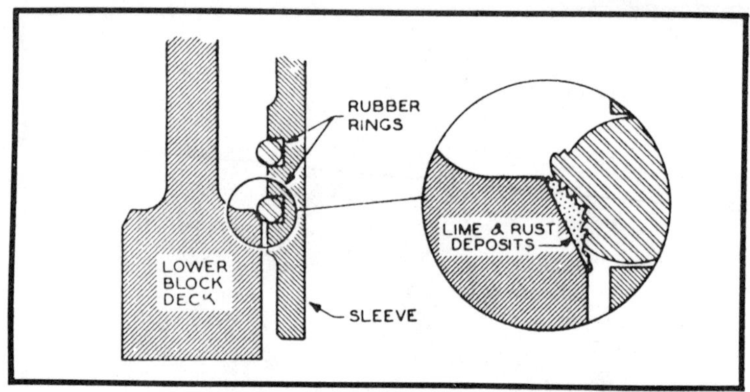

Fig. 6-45. Remove lime and rust deposits on the cylinder flanges. If necessary, these flanges can be machined. (Courtesy Sealed Power Corp.)

Carefully clean rust and scale deposits from the flange areas that would interfere with gasket sealing (Fig. 6-45). Use new gaskets, all from the same manufacturer who made the liner. There are slight differences in gasket and groove dimensions between original equipment and replacement parts that can be disastrous when the parts are mixed. Mount the gaskets on the liner, lubricate with hydraulic fluid or green soap, and slowly lower the liner into the block. The gaskets must not be allowed to turn as the liner goes home. Use the mold flash, the line on the ring's outer diameter, as the guide. A twisted gasket will refuse to displace and bulge the liner inward against the piston.

Deck height varies with make and model in such a way that generalizations would be fatal. Check the engine manual. Figure 6-46 shows how the liner is compressed with shop-fabricated clamps mounted over the cylinder bolt holes and torqued to specification. Because of the forces involved and the need for accurate torque readings, the clamps are made from cold-rolled steel and the washers are ground and hardened. Add or subtract shims at the upper flange to establish the correct deck height.

Dry-sleeve engines are easier to work, since there is less to go wrong. Inspect the flange at the base of the block for evidence of fret wear. In most cases a reputable machinist can dress the flange and compensate for the metal lost with shims. Clean the block bores and lightly sand to remove carbonized oil, surface rust, and scores that would hinder installation. Linear fit is subject to some dispute.

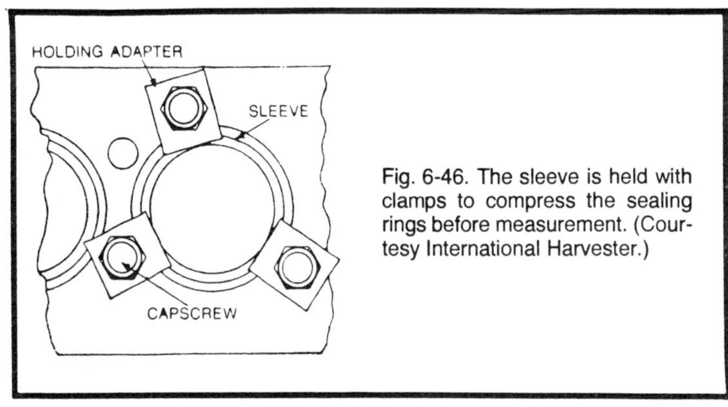

Fig. 6-46. The sleeve is held with clamps to compress the sealing rings before measurement. (Courtesy International Harvester.)

Aftermarket suppliers have sometimes argued for a light fit, one that would allow the liners to be installed three-quarters of the way with palm pressure. Engine makers generally suggest that the liner be pressed home with a force on the order of 5000 lb. Figure 6-47 shows an installation tool intended to distribute these forces over a wide area.

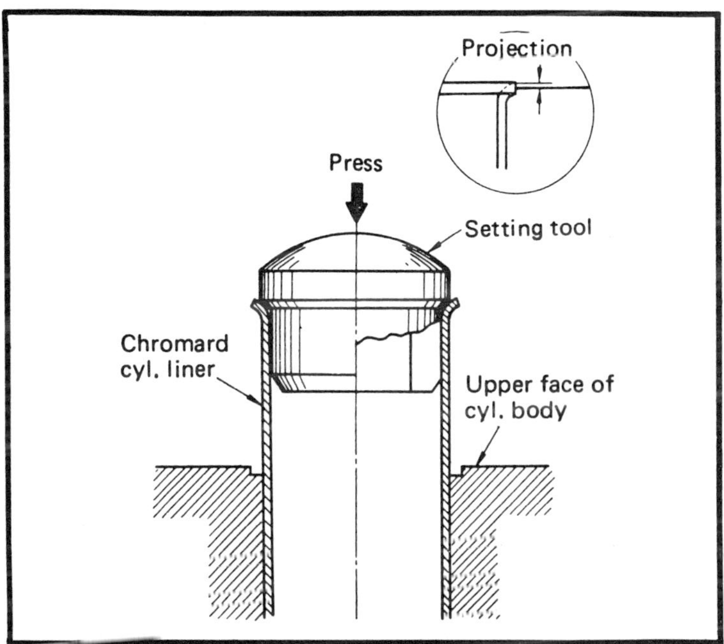

Fig. 6-47. An arbor is available from Isuzu to install the sleeve without distorting it.

Camshafts

Figure 6-48 illustrates the camshaft assembly for the Isuzu series 6BB1 engine. As with all other in-block designs, the cam bearings are pressed into bosses in the block and are lubricated with oil from the main gallery. Overhead camshafts generally ride on split bearings which are dismantled for camshaft removal. Bearing caps and leading edges must be marked for correct assembly.

The first step is to measure camshaft float, or endplay. Insert a feeler gauge between the block and drive gear or drive sprocket. Using a soft-faced mallet, tap the cam back toward the flywheel. Next, measure the distance between block and drive gear or the sprocket, as the case may be, with a feeler gauge. If you use a dial indicator, its plunger should rest on the face of the wheel and the dial be on zero. With a screwdriver, pry the gear forward to the full extent of cam movement. The amount of movement is camshaft float, which should not exceed 0.09 in. for most engines. Excessive float can be corrected by replacing the thrust plate, a part referenced in Fig. 6-48 as No. 4. If the maker was so thoughtful as to supply a symmetrical plate, it may be turned over to return camshaft float to specs.

Nearly all tappets are mushroom-shaped and locked in their compartment until the cam is removed. However, if the engine is upright, the tappets will fall down and bind against the cam journals, blocking camshaft extraction. The obvious solution is to turn the engine over, so the tappets drop clear. If this is not possible, as for example in a case where the engine is installed in the vehicle, the tappets can be secured with patent spring clips, available from K-D Tools, Snap-On, and other tool makers.

Withdraw the camshaft, being careful not to bang the journals against the bearings. Inspect the wear patterns on the lobes. On nearly all engines, and I believe on all that are currently in production, the lobes bear on one side of the tappets, giving them a twist as they rise (Fig. 6-49). If there is center contact, the camshaft and tappets will soon fail, and their failure will be dramatic: once the case-hardening is worn through, the tappet cups and the lobe disintegrate. Lobe/tappet pressure is 200,000 psi at idle.

As a further check, compare lobe and journal diameters against factory wear limits.

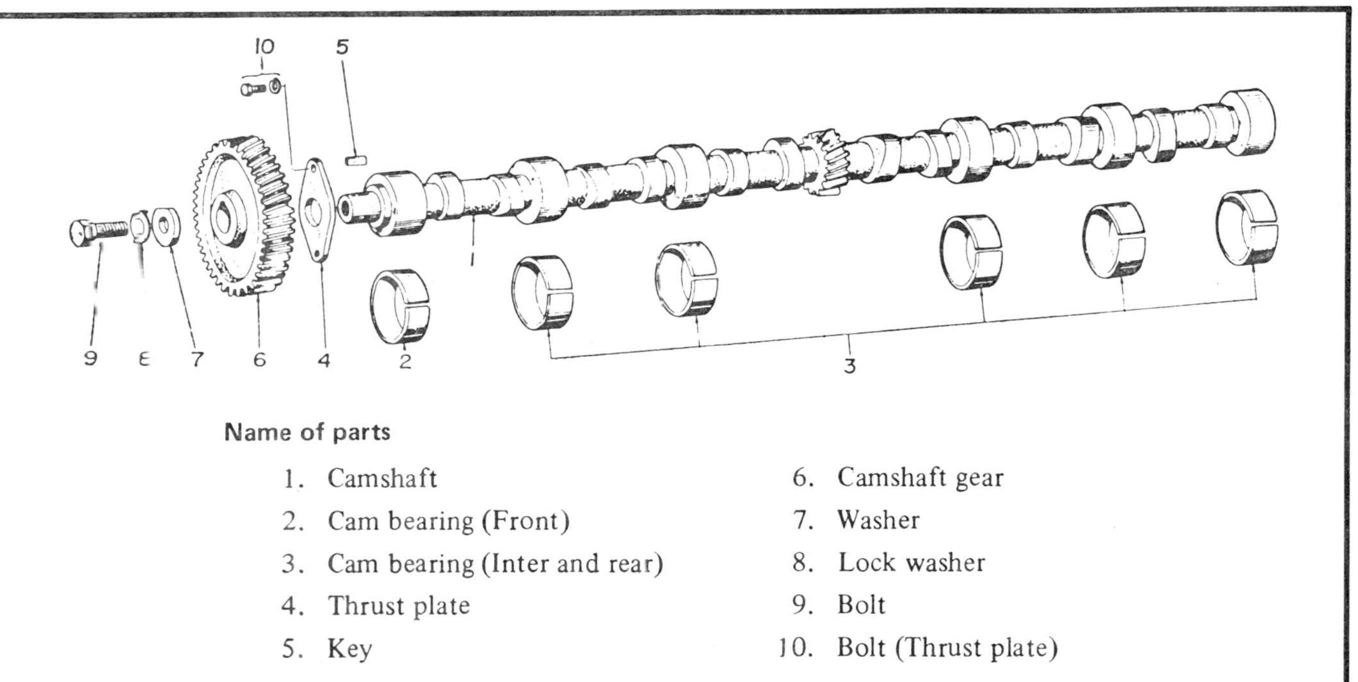

Name of parts

1. Camshaft
2. Cam bearing (Front)
3. Cam bearing (Inter and rear)
4. Thrust plate
5. Key
6. Camshaft gear
7. Washer
8. Lock washer
9. Bolt
10. Bolt (Thrust plate)

Fig. 6-48. A typical camshaft assembly. (Courtesy Isuzu Motors, Ltd.)

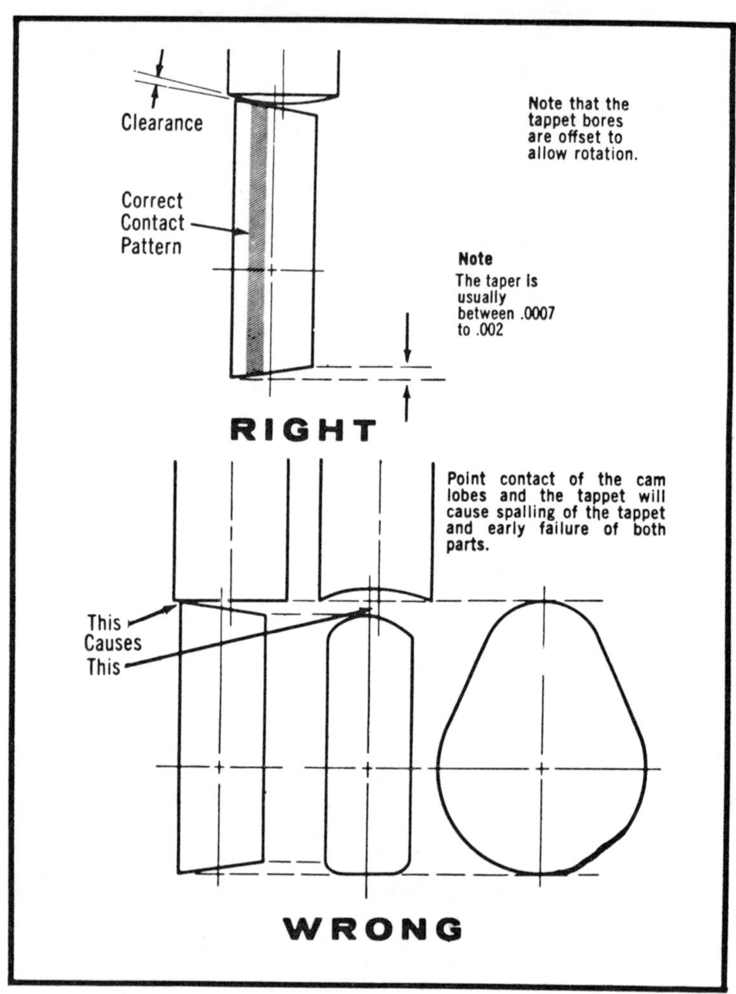

Fig. 6-49. Most cams are ground on a slight taper and ride eccentrically on the tappets to provide a scrubbing action for valve rotation. Wear rounds the cam lobes and dishes the tappets. (Courtesy Sealed Power Corp.)

Bent camshafts are by no means unusual. Manufacturing tolerances allow 0.02 in. deflection at each end when measured from the center journal. This specification can be easily met by an automotive machinist with the appropriate equipment.

Camshaft Bearings

Inspect the bearings for scores and other surface irregularities. It is prudent to have the bearings miked. While an engine will

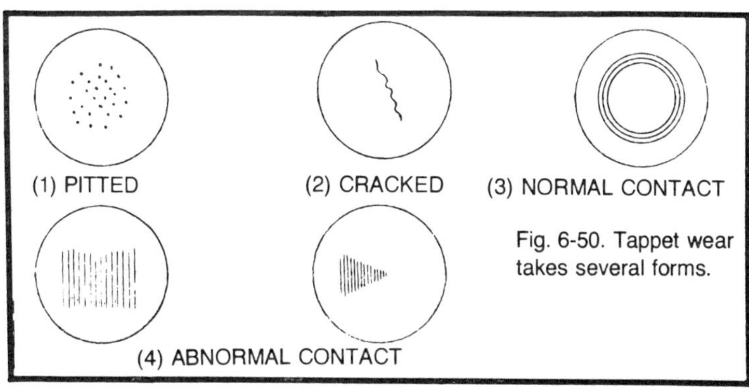

Fig. 6-50. Tappet wear takes several forms.

tolerate a rattling camshaft, excessive bearing clearance can increase lube oil consumption. Oil spills out of the bearings and down over the crankshaft where it is flung onto the cylinder walls.

Bearing replacement requires the use of a special tool. The only precaution necessary is aligning the oil holes, especially in the case of Detroit Diesel, Isuzu, and a few other engines where a split bearing will affect the oil-supply ports and the webs.

As you can appreciate from the extreme pressures involved, camshaft lubrication is critical. The bearings are served by coating them with lube oil prior to crankshaft installation. Lobes and tappets should be greased with EP lubricant, sometimes sold in cannisters as assembly lube. New camshafts are already lubricated and should be installed as received from the factory.

Tappets

As far as I can tell, all diesel engines suitable for automotive use run mechanical, or solid, tappets. The occasional need for valve lash adjustment is accepted for the reliability afforded by mechanical tappets.

If they are to be reused, rack the tappets in numbered pigeonholes. Examine the flanks for scores, which would indicate a failed oil filter or simple neglect to change the lube oil, and check the tappet feet for wear. Acceptable and abnormal wear patterns are shown in Fig. 6-50. If any tappet is worn, you should replace the camshaft and all of the tappets. Total replacement of parts is initially expensive, but will save money over the long run. Tappets and camshaft are like meshed gears, in that replacing one dooms the other.

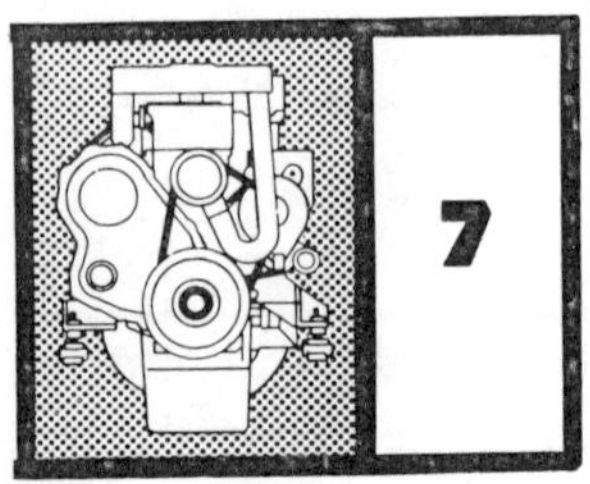

7 Installing The Engine

At this point any work on a used engine should be finished. The engine should have been overhauled and its pumps and injectors calibrated (Fig. 7-1). A new engine can be installed from the crate, although a careful mechanic will drop the pan to clean out the swarf and sand that accumulates in a new engine.

BELL HOUSING/TRANSMISSION

The bell housing/transmission nexus can be the most critical problem in the swap. There are three alternatives:

1. You can use an engine that bolts up to the transmission.
2. You can use an engine/transmission combination, replacing the original box.
3. You can make a bell housing/transmission adapter plate.

The first option limits engine choice, since few engines other than the Chrysler-Nissan are sized to mate with American transmissions. Isuzu engines are adapted by the importer, DeLoran Diesel of Livonia, Michigan.

The second option involves a used automotive engine and transmission combo. For practical purposes this means Mercedes-Benz, because you might as well look for a Dusenberg as for a Peugeot in a junkyard. A minor problem encountered with Mercedes

Fig. 7-1. Out with the old, in with the new.

drivelines is the absence of a universal joint. The differential is bolted to the undercarriage and drives the wheels through half-shafts (Fig. 7-2). Although some movement is allowed by rubber grommets and some chassis flex can be expected, the differential pod is essentially fixed. The forward universal joint is replaced by a rubberized coupling (Fig 7-3).

The swapper has three alternatives. The best, from the point of view of ride-comfort, silence, and systems integrity is to swap the

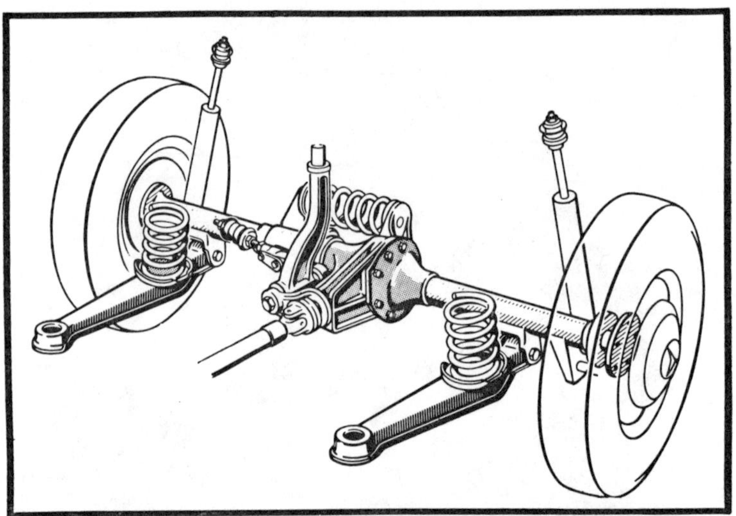

Fig. 7-2. Mercedes-Benz addiction to swing axles can be a problem for swappers.

whole driveline, including engine, transmission, and third member. Problems occur down the line at the rear suspension mounts. You'll have to accommodate the swing arms, bracketing them to withstand acceleration and braking loads. Spring perches and shock mounts will have to be fabricated, as well as a mount to contain the torsional forces generated by the differential. Track, or the distance between

Fig. 7-3. Instead of a U-joint Mercedes transmissions drive through a rubber doughnut.

wheel centers, is 57.4 inches (1460 mm) on the 180 and 190, and 57 7/8 inches (1470 mm) on the 220. Narrowing the track to fit a smaller chassis is possible but expensive.

The second alternative would be to replace the transmission flange with a U-joint so that the vehicle's original third member could be retained. This option would involve extensive machine work and, so far as I know, is not done.

The third alternative is the least venturesome of the three. Some driveshaft work is required, but none of it is out of the ordinary for a street-wise speed shop.

The original third member is retained, connected to the transmission by a two-piece drive shaft. Once the engine and transmission are located in the chassis, you will have to:

1. Mate the Mercedes driveshaft-to-transmission flange to a short (12-18 in.) length of shafting. Most swappers use the forward shaft from a truck (Fig. 7-4).

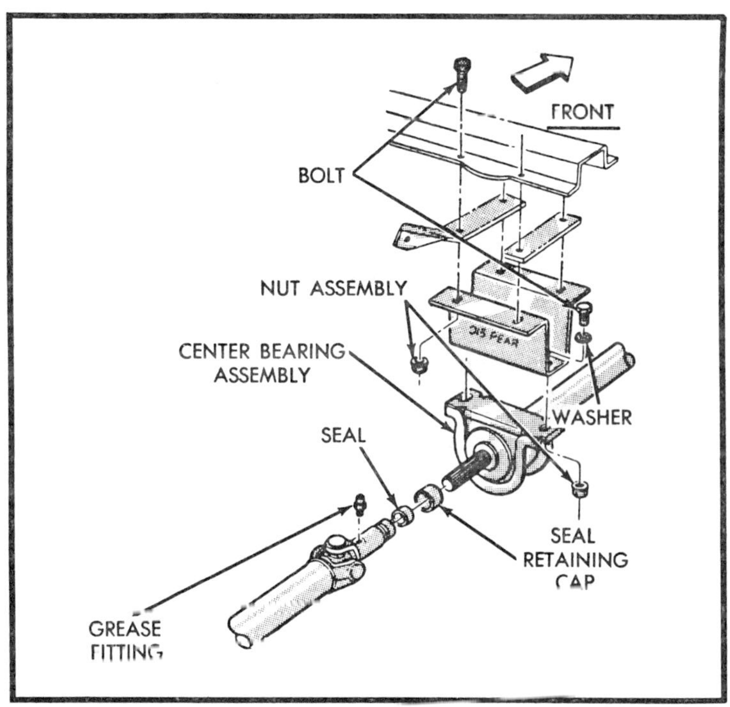

Fig. 7-4. Some passenger cars and all trucks employ a driveshaft center bearing. (Courtesy Chrysler Corp.)

2. Mount a center bearing to support this shaft, one end of which terminates in the Mercedes flange, the other which is splined to allow U-joint movement.
3. Carefully measure the distance between the splined end of the forward shaft and the U-joint yoke at the differential. If you are concerned about the accuracy of this measurement—and it must be dead on—have an experienced person make it.
4. Have the rearward shaft cut, fitted with a U-joint compatible with the differential yoke, and balanced.
5. Assemble the drivetrain.

The machine work and welding must be handled by a professional. You are responsible for driveshaft measurement and the geometry of the driveline. The driveshaft splines, which are just aft of the center bearing, must not bottom when the rear suspension is at the upper limit of its travel. The rubber snubbers between the frame and axle and not the splines should define the upper limit of suspension travel. By the same token, there should be at least 5/8 in. of spline/yoke engagement with the suspension on full rebound. Spline engagement is determined by shaft length and some small correction can be made by moving the engine and transmission assembly forward or aft.

The hanger should be mounted on a frame cross member and the bolt holes should be elongated to allow adjustment. The centerline of the shaft should pass through the center of the bearing, so that the rubber coupling will be unloaded during normal operation. Position the hanger, put the transmission in neutral, and turn the shaft by hand to detect binds. Move the hanger laterally in the elongated bolt holes or vertically by means of shims between it and the cross member.

While both U-joint yokes should be within 1 degree of being parallel, the driveshaft should be angled downward between 7 and 9 degrees (Fig. 7-5). A lesser angle would be insufficient to rotate the U-joint bearings and a greater angle would work them to early failure.

The angle is determined by the height of the engine, for it is impractical to attempt much in the way of rear suspension modification. The engine can be lifted as a unit with the transmission, or it can

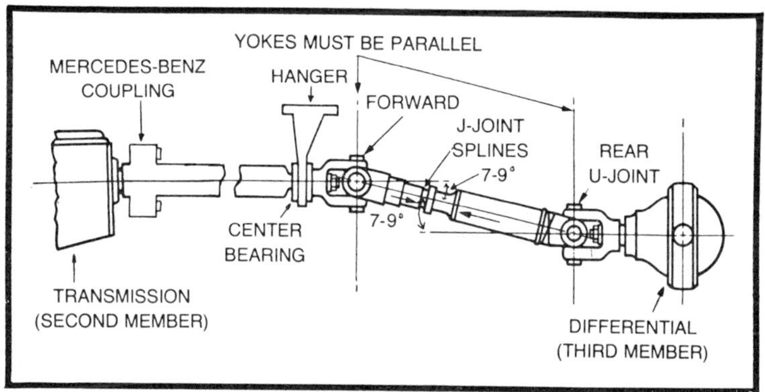

Fig. 7-5. Some driveshaft angle is needed to exercise the U-joint bearings. (Courtesy Chrysler Corp.)

be tilted upward at the front motor mounts. Diesel engines do not, of course, have a carburetor with a float and so they are not sensitive to installation angles. Some engines can even be tilted 30 degrees without interfering with the lube oil supply in the sump, although no one would want a conversion with the fan pointing skyward.

Adapter Plates

For practical purposes there are two kinds of adapters, both made from steel or aluminum plate. The plate is made of steel or aircraft-quality aluminum and should be blanchard ground on both sides.

If the engine will not bolt up to the original transmission and if you do not have a transmission for the engine, then you will have to construct an adapter plate. People tend to have exaggerated respect for adaptor plates, believing that it is impossible to mate an engine to a foreign transmission without placing a master machinist on retainer. Actually the work is simple, at worst requiring two fairly elementary machining operations. The rest of it you can do yourself.

Bellhousing-to-Transmission

This fairly rare adaptation takes the bellhousing supplied with the engine and mates it with the vehicle's original transmission. Some engine makers supply an automotive bellhousing for this purpose, but other manufacturers lock the swapper into their own bellhousing because of starter interference or clutch-diameter limitations.

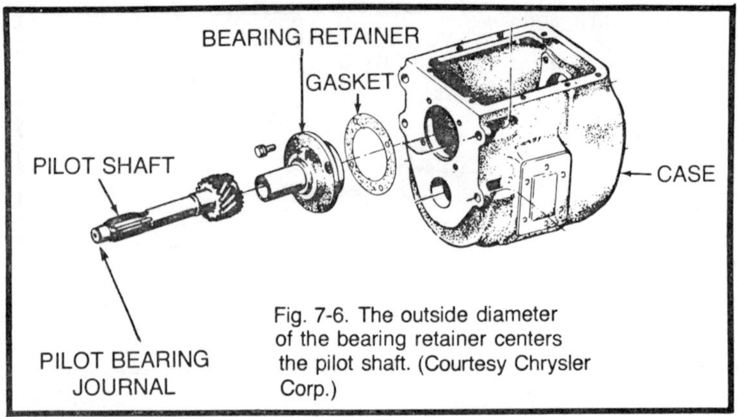

Fig. 7-6. The outside diameter of the bearing retainer centers the pilot shaft. (Courtesy Chrysler Corp.)

The first requirement is to find the center of the pilot shaft. Fortunately, transmission designers have simplied things for the swapper and for their other customers by centering the pilot shaft on a bearing flange. This part, labeled "bearing retainer" in Fig. 7-6, is usually known as the pilot. Once its center is found, the rest of the job is easy.

Remove the pilot and take it, together with the plate, to a machine shop. Explain that you want a palm-push fit between the two parts.

The forward end of the pilot shaft rides on a bearing in the crankshaft flange or, on a few diesels, at the flywheel hub (Fig. 7-7). Chances are that the original bearing will not be sized correctly for the pilot shaft. The bushing, or, as it may be, needle bearing, can be extracted with a small puller. If such a tool is not available, remove the bushing by packing the cavity with heavy grease. Obtain a rod with the same diameter as the original pilot. Use it as a hydraulic ram, hammering it into the grease. Hydraulic pressure will force the bearings back and out. With needle bearings use grease-soaked newspaper shreds as the working medium.

·Surprisingly, often you will be able to purchase the correct replacement from a bearing-supply house. If not, try to find a bushing that is correct in one dimension and machine the other. It is unnecessary to use a needle bearing here. A sleeve can be placed between the bearing outer diameter and the boss. Because the shaft will retract out of the bearing by the thickness of the adapter plate, it is useful to keep this dimension as small as possible for adequate bearing surface and avoidance of clutch-linkage problems. A 5/8 in. aluminum plate is the usual choice, although some builders have used 1/2 in. steel without trouble.

Once we have the pilot hole cut in the adapter and the pilot bushing sized, the rest of the work will be anticlimactic. Mount the transmission on the plate and mate the output shaft with the pilot bearing. Mark the bolt holes and drill.

Bellhousing-to-Engine

Retaining the vehicle's original bellhousing is usually wiser than trying to adapt the engine's bellhousing to the transmission. However, automatic and a few foreign manual boxes with integral bellhousings allow no choice in the matter: the drive train nexus must be at the bellhousing and block.

All dimensions derive from the center of the pilot shaft. Size the pilot shaft to the flywheel-flange bushing as indicated in the previous section. The thickness of the adapter plate is normally between 1/2 and 5/8 in., although the length and bearing penetration of the pilot shaft will affect this.

Cut the center out of the plate for flywheel clearance, making provision for the starter nose. Sandwich the plate between the block and bellhousing/transmission assembly. Have a helper hold the

Fig 7-7. The bushing in the center of this scarffed-up flywheel supports the forward end of the pilot shaft.

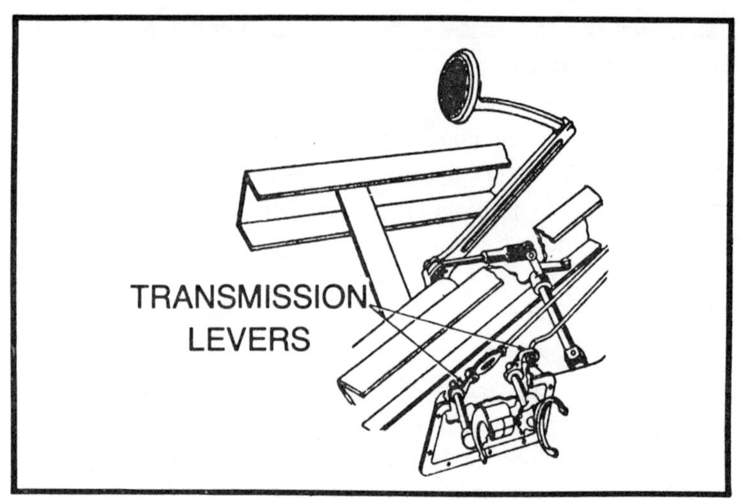

Fig. 7-8. Old Fords never die, but swappers don't need to emulate the clutch linkage on this 1940 example.

assembly hard against the plate with the pilot shaft engaged in the flange bushing. Carefully mark the centers of the bolt holes and drill them with a press. Bolt the assembly together and scribe an outline of the bellhousing on the plate. As before, trim the plate with a band or saber saw.

Clutch Linkage

As I said earlier, the basic rule of engine swaps is to respect systems integrity. The clutch linkage is no exception. Retain the original, unless you are absolutely forced to do otherwise. However, changing bellhousings or substituting a manual box for an automatic will mean that you will have to modify the existing linkage or fabricate a new one.

The clutch can be disengaged in any of three ways. A system of leverage-multiplying rods and levers was traditionally used (Fig. 7-8). Reliable rod-and-lever articulation is not as simple as it looks and can make embarrassing space demands. Another approach, used on many modern cars, is cable articulation (Fig. 7-9). The mechanism is self-contained with anchoring points at the firewall and bellhousing. The flexibility of the cable simplifies the installation.

The ultimate solution, favored by foreign manufacturers whose vehicles must be adaptable to left- and right-hand drive markets, is a hydraulically actuated clutch. Figure 7-10 is a cross section of a Peugeot clutch and slave cylinder. Other designs employ a threaded piston rod for clutch adjustment. Figure 7-11 is a cutaway of a typical master cylinder that is mounted to the firewall, forward of the clutch pedal. The two are in no fixed spacial relationship to each other and the flexible connecting line can be routed in whatever manner is convenient.

Cross Members

The swap may involve a new front or rear cross member. Study the originals carefully before you dismiss them out of hand: sometimes a small modification will salvage them. If not, scout the junkyards for a suitable replacement or build one yourself.

Rear cross members are simple constructs that can be duplicated with 1/4 in. steel channel or 0.125 in. seamless tubing (Fig. 7-12). Tubing has a far better strength-to-weight ratio than channel and can be bent to specificaiton at a muffler shop.

A front cross member that merely supports the engine can be moved, modified, or replaced at will. However, if it serves as the tiepoint for the suspension, it should not be touched without a degree in automotive engineering (Fig. 7-13). Unfortunately, most

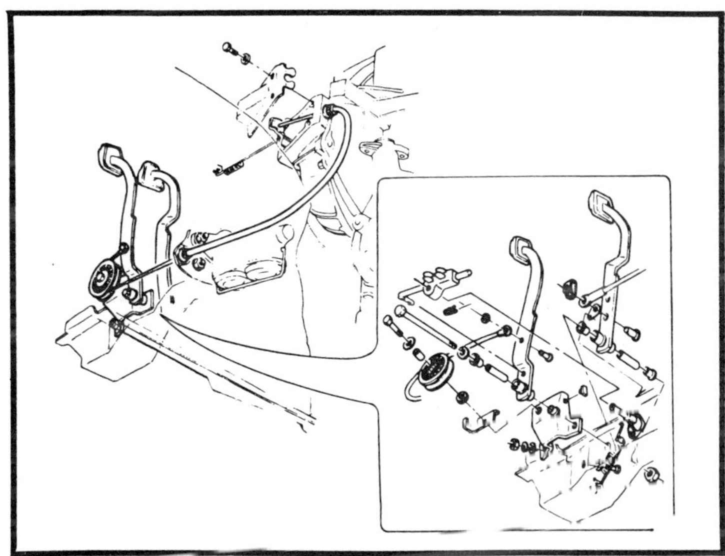

Fig. 7-9. Chevrolet and most other modern cars employ a clutch cable.

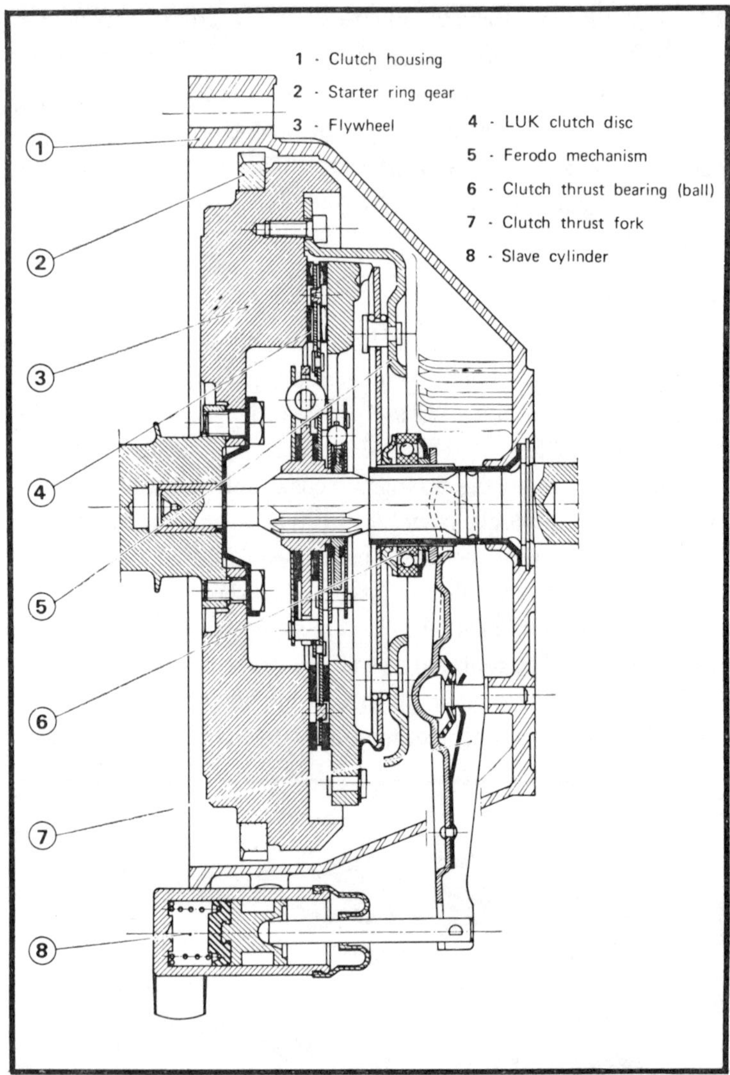

Fig. 7-10. In most installations the piston diameter of the slave cylinder is the same as the master cylinder. Increasing the bore of the slave cylinder would reduce the stroke and cut pedal effort.

modern vehicles, particularly those with Macpherson suspensions, employ these multiduty cross members. Sometimes you can get around the problem by trimming the engine's pan for clearance and constructing a second cross member, aft of the first, to bear engine weight.

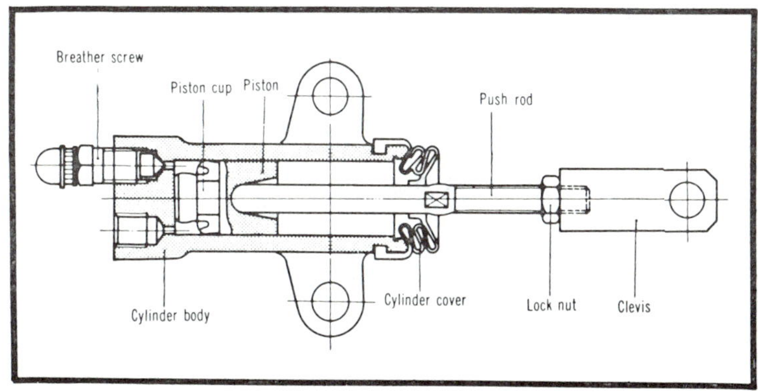

Fig. 7-11. A Mitsubishi master cylinder. The clevis is threaded for pedal-height adjustment.

Motor Mounts

Lower the engine/transmission assembly into place, bolting the transmission if the original mount can be used. If not, connect the drive shaft to aid in alignment. Supporting the engine with the hoist, center it in the compartment. Normally the engine is mounted dead horizontal unless there is a driveline problem, which complication was discussed earlier in this chapter.

Motor mounts have three elements: the engine bracket, the rubber insulator, and the frame bracket. The frame bracket may be

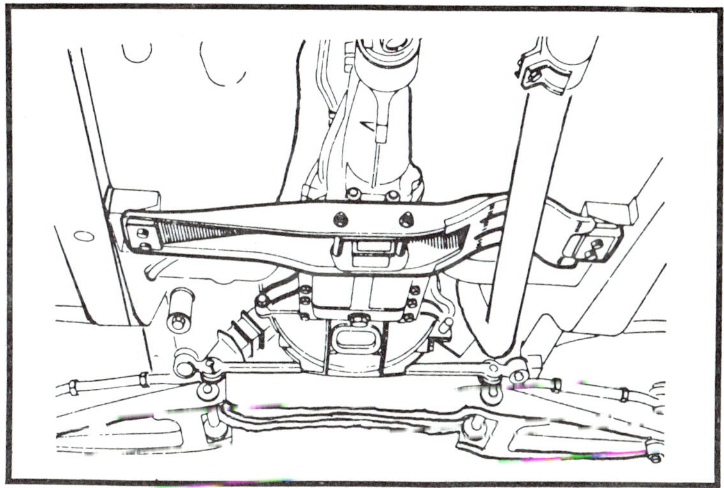

Fig. 7-12. A rear cross member is simple enough to fabricate. (Courtesy Chrysler Corp.)

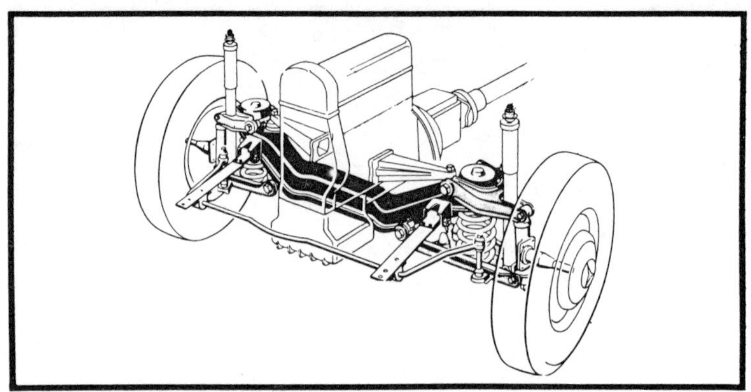

Fig. 7-13. Mercedes-Benz, as well as most other manufacturers, hangs the suspension from the cross member.

integral with the chassis or it may be bolted into place (Fig. 7-14). The width of the engine bracket is determined by the distance between the bolt holes in the block, but the width of the chassis bracket is usually up to you, since it is doubtful that an existing integral bracket could be used in an swap. Make the bracket as broad-footed as is consistent with common sense. The wider the bolt spacing is, the lower the unit-area loads will be and the less opportunity for fatigue cracks.

Another consideration is the direction that loads will take through the mounts. Ideally the chassis bracket should stand vertical off the frame or cross member. The more it is tilted, the greater the tendency of the chassis to flex inward under engine loads. Cantilever the mounts as necessary; that is, extend the engine brackets out at right angles to the block while the chassis brackets rise vertically.

As found, industrial and marine engines are supported well forward at points near the timing case. While this is not in itself objectionable, it does tend to load the rear mount. These same blocks have more than one set of drilled and tapped engine bracket bolt holes, so there is some choice in the matter.

Purchase new insulator biscuits, even if you are using the type found on the car. As humble as they look, insulator biscuits are the beneficiaries of much computer time. It is worth the trouble to choose a set intended for diesel passenger cars or light trucks. As for the brackets, they can't be too strong. At a minimum, use 0.125 in. wall tubing or 1/4 in. steel plate. Bolts should be SAE grade 8 or grade 6. The latter are available from automotive parts houses, the former from Caterpillar dealers.

Meanwhile the engine is still supported on the hoist. We need to construct cardboard patterns of the brackets, a much more accurate procedure than transferring measurements directly to steel. Incorporate the biscuits into the patterns.

The boltholes for the chassis brackets should be elongated to allow for some small range of adjustment. Unless the biscuits are horizontal, the center hole in the engine brackets—the hole for the biscuit stud—must also be elongated. A look at Fig. 7-14 should make things clearer. There must be room for the engine bracket to settle on the biscuits. Unless the holes are elongated, both bracket and engine will hang on the studs.

Once the engine is bedded, the rest of the swap will consist of detail work as outlined in an earlier chapter. Most of this work is opportunistic, doing what you have to do with what you have; so talking about it is not very helpful. But there are a few points worthy of mention.

Wiring

Some changes to the wiring are needed to accommodate diesel power. The alternator and voltage regulator must be matched; that

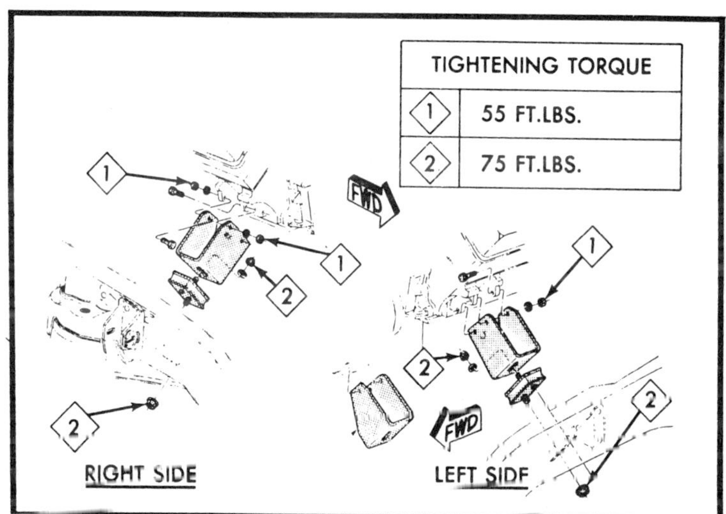

Fig. 7-14. Frame brackets are hidden by the chassis rail and represented by dotted lines. (Courtesy Chrysler Corp.)

is, if you use an alternator that came with the engine, you cannot use the vehicle's voltage regulator. Oil pressure and coolant temperature senders are usually compatible with dashboard gauges, although not necessarily with idiot lights. In any event, aftermarket gauges and sensors can be used. The battery capacity should be increased, particularly in cold climates.

Diesel starters are gasoline-engine starters on a larger scale. The internal wiring and control circuitry is no different. Connecting the "S" lead from the ignition switch to the starter solenoid should engage it. The "I" (ignition) lead is redundant and can be discarded. The engine can be stopped manually by means of a Bowden cable

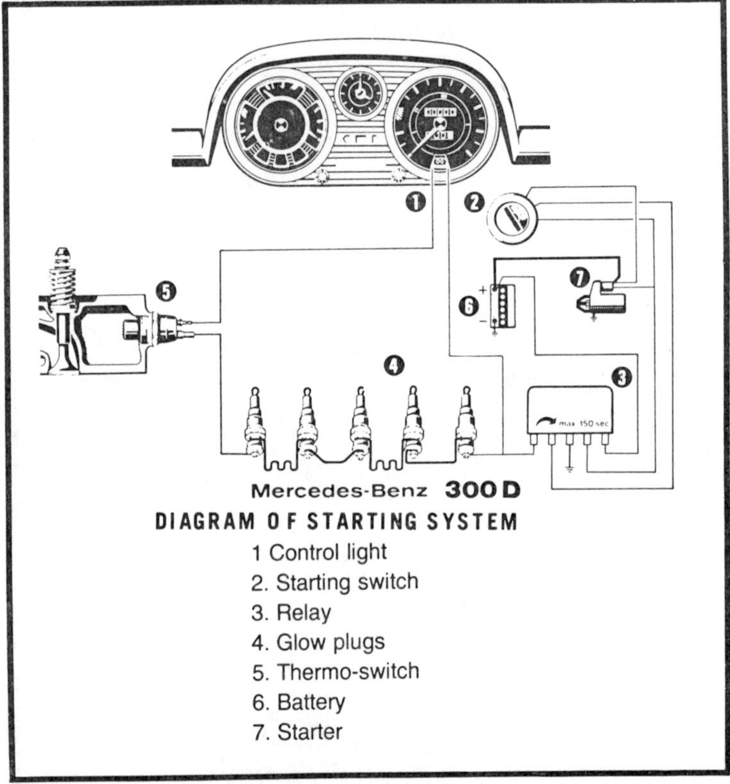

Fig. 7-15. Mercedes-Benz glow plugs can be switched on when the starter is engaged, if the thermoswitch 5 senses that the coolant is cold. The time-controlled relay 3 automatically opens the circuit. Swappers eliminate the complexity (and convenience) of this system by wiring the relay through a push-button switch. As long as the switch is held down, the circuit from battery through relay to heater plugs is complete.

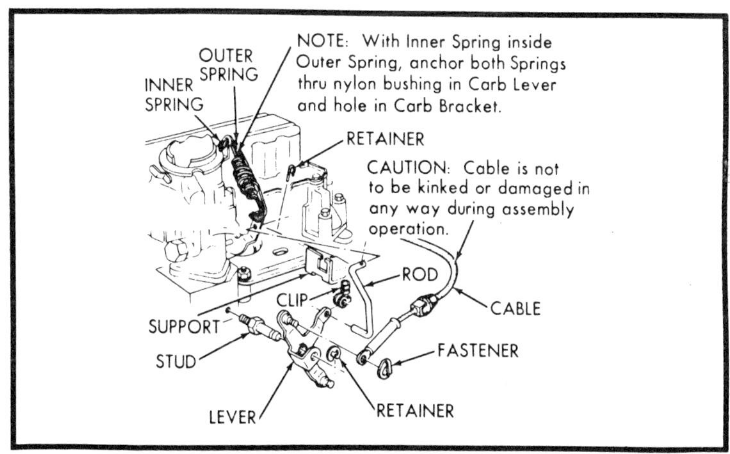

Fig. 7-16. A cable-operated throttle, together with dire reminders of things that can go wrong. On a diesel the free end of the cable would attach to the fuel pump throttle lever (centrifugal governor) or to the venturi control lever (vacuum governor). (Courtesy General Motors Corp.)

connected to the stop lever on the fuel pump, or this function can be integrated into the ignition switch. Rig the stop lever to work by means of a solenoid and wire the solenoid to the "A" (accessory) terminal of the switch. Turning the key to the right engages the starter; turning it full left stops the engine.

Heater plugs are wired in parallel to each other (Fig. 7-15). If one should fail, the others will continue to function. Because of the heavy current draw, the plugs are fed through a relay mounted on the firewall or inner fender. The relay is triggered from the inside of the passenger compartment by a normally-open switch.

Throttle Control

The throttle control is a safety-related item that keeps engineers awake at night. Spend some time engineering yours. Purchase the complete mechanism—cable, bracket, and all—from a junkyard and replace the cable if it operates stiffly or shows core separation. The warnings printed in Fig. 7-16 are General Motors' way of reminding you what can go wrong with OEM installations.

When all the work is done, and you have driven your creation around the block, you deserve congratulations for a job well done. May you have many low-cost, low-maintenance miles!

Index